Sustainable Civil Infrastructures

Editor-in-chief

Hany Farouk Shehata, Cairo, Egypt

Advisory Board

Khalid M. ElZahaby, Giza, Egypt
Dar Hao Chen, Austin, USA

Steering Editorial Committee

Dar Hao Chen, Texas A&M University, USA
Jia-Ruey Chang, National Ilan University, Taiwan
Hadi Khabbaz, University of Technology Sydney, Australia
Shih-Huang Chen, National Central University, Taiwan
Jinfeng Wang, Zhejiang University, China

Wissem Frikha · Shima Kawamura
Wen-Cheng Liao

Editors

New Developments in Soil Characterization and Soil Stability

Proceedings of the 5th GeoChina International Conference 2018 – Civil Infrastructures Confronting Severe Weathers and Climate Changes: From Failure to Sustainability, held on July 23 to 25, 2018 in HangZhou, China

 Springer

Editors
Wissem Frikha
Department of Civil Engineering,
  National Engineering School of Tunis
University of Tunis El Manar
Tunis, Tunisia

Wen-Cheng Liao
Department of Civil Engineering
National Taiwan University
Taipei, Taiwan

Shima Kawamura
Graduate School of Engineering
Muroran Institute of Technology
Muroran, Japan

ISSN 2366-3405   ISSN 2366-3413 (electronic)
Sustainable Civil Infrastructures
ISBN 978-3-319-95755-5   ISBN 978-3-319-95756-2 (eBook)
https://doi.org/10.1007/978-3-319-95756-2

Library of Congress Control Number: 2018948647

Printed on acid-free paper

This Springer imprint is published by the registered company Springer International Publishing AG part of Springer Nature
The registered company address is: Gewerbestrasse 11, 6330 Cham, Switzerland

Contents

Introduction

The geotechnical engineering is a relatively new discipline. Its importance is closely related to the cost of all types of buildings and constructions. The evolution of theories and investigations looks to represent the real state of the soil behaviour under static load or during complex configuration by the use of 2D or 3D numerical models. The employed parameters taken from soil investigations should be chosen as correct as possible to avoid any design errors. It is then indispensable to respect the evolution of the different techniques of soil characterization, to improve them and to look for innovative solutions to ensure the ground stability.

This volume is part of the Proceedings of the 5th GeoChina International Conference 2018 – Civil Infrastructures Confronting Severe Weathers and Climate Changes: From Failure to Sustainability, held on July 23 to 25, 2018 in HangZhou, China. The proceedings of this volume present some responses on soil characterization and stability due to climate changes. This conference has been endorsed by a number of leading international professional organizations and provides a showcase for recent developments and advancements in design, construction and safety inspections of transportation infrastructures. Conference topics cover a broad array of contemporary issues for professionals involved in bridge, pavements, geotechnical, tunnel and railway engineering, as well as emerging techniques for safety inspections.

The papers were reviewed by the organizing and scientific committees. All papers were classified by theme and reviewed following the same technical standards. A minimum of two full reviews coordinated by various track chairs was undertaken and supervised by the editors of the volume. The authors of the accepted papers have addressed all the comments of the reviewers to the satisfaction of the track chairs, the volumes editors and the proceedings editor. Publication of this quality of technical papers would not have been possible without the dedication and professionalism of the anonymous papers reviewers.

As a result, the present volume includes a widespread of relevant 11 papers covered by the subjects of the soil characterization, on the one hand, and the soil stability, on the other hand. The editors wish to thank the authors for their valuable contributions in the preparation of the papers, the organizing committee and the scientific committee for their full engagement in making this publication possible. Thanks are also extended to Springer for their coordination and enthusiastic support to this conference.

Impact of Liquid Whey Waste on Strength and Stiffness of Cement Treated Clay

Thang Pham Ngoc[(✉)], Behzad Fatahi, and Hadi Khabbaz

School of Civil and Environmental Engineering, University of Technology
Sydney (UTS), Sydney, NSW, Australia
Thang.PhamNgoc@student.uts.edu.au,
{behzad.Fatahi,hadi.khabbaz}@uts.edu.au

Abstract. The reuse of whey waste, a by-product of the dairy industry, is an emerging issue due to the environmental impacts. Some previous experimental studies have indicated that whey waste can be used as an admixture for cement-based materials, including mortar and concrete, to reduce the setting time and increase the workability, thus reduce the amount of required cement. However, influence of whey waste on cemented soil has not received sufficient attention. This study investigates variations of unconfined compressive strength (UCS) and Young's modulus (E) of cemented Kaolin clay when water in cement slurry was replaced by different whey waste proportions. Unconfined compression tests were conducted on treated specimens after two different curing times, namely 14 days and 56 days. Stress-strain relationship in each test was used to compute UCS and E at different dosages of cement and whey waste. Results of the experiments show improvements of UCS and E only for specimens when less than 10% water in cement slurry was replaced by liquid whey waste at 56 day-curing age, regardless of cement dosage. For the other cases, the presence of whey waste resulted in reductions of both UCS and E, indicating that although whey waste can be used to improve mechanical properties of cement treated clay, the optimum dosage should be selected very carefully to minimize the adverse effects. Different responses of UCS and E with curing age, dosages of cement and liquid whey waste are explained while discussing about the effects of lactose (milk sugar) available in whey waste acting as a retarding agent.

1 Introduction

Whey waste is a by-product of the cheese production. Liquid whey waste typically consists of about 93% water, 5% lactose and smaller quantities of fat, protein and minerals (Kelling and Peterson 1981; Modler 1987). It has commonly been considered as a waste product due to low concentration of milk constituents (Siso 1996). The liquid fraction of whey is drained from the curd during the manufacture of cheese. Typically, every 100 kg of milk gives about 15 kg of cheese, depending on the variety, and about 85 kg of whey. There are two categories of whey, sweet and acid whey, which are by products of cheddar cheese and cottage cheese manufacturing, respectively. Whey waste is reported as the main pollutant generated from the process of making cheese. If whey waste was directly discharged to the environment, it would harm the receiving environments by inducing an excess of oxygen consumption,

© Springer International Publishing AG, part of Springer Nature 2019
W. Frikha et al. (eds.), *New Developments in Soil Characterization and Soil Stability*,
Sustainable Civil Infrastructures, https://doi.org/10.1007/978-3-319-95756-2_1

toxicity, etc. (Marwaha and Kennedy 1988; Palmieri et al. 2017). Whey waste biological oxygen demand (BOD) is 32,000 to 40,000 ppm, which creates a very severe disposal problem.

A number of studies have been conducted in an attempt to reuse whey waste. Animal feeding trials conducted by McIntosh et al. (1998) showed that whey protein can be used as functional food ingredients in retarding colon cancer. Cuzman et al. (2015) reported that whey waste can be used as a potential nutrient source for biotechnological use of Sporosarcina pasteurii, a bacterium with the ability to produce biological cementation and has potential use in ground improvement and in developing bio-materials for the building industry. Results of experiment by Taube et al. (1974) revealed that the use of whey waste as an admixture in mortar and concrete reduces the setting time and increases the workability. A saving of at least 5% cement was reported. Effects of whey waste on chemical and physical properties of soil have also been investigated in a large amount of studies (Lehrsch et al. 2008; Robbins and Lehrsch 1998; Sharratt et al. 1962). On the contrary, effects of whey waste on mechanical properties of soil, which may bring in potential benefits in soil improvement, has received little attention.

The main objective of this study is to conduct experimental investigation on influence of whey waste on strength and stiffness of a cement treated clay. Variations in mechanical properties of treated soil were expected during a series of unconfined compression tests on cement treated kaolin when water in cement slurry was partially to fully replaced by liquid whey waste.

2 Laboratory Experiment

2.1 Sample Preparation

Commonly available kaolin Q38 clay was used as the base material treated by cement and whey waste. General purpose cement was selected due to the popularity in construction, while liquid whey waste was provided by Paesanella, a cheese manufacturer in New South Wales, Australia.

All mixing processes in the study were performed manually. The soil mixture, consisting of kaolin, cement, water and whey waste, was prepared by mixing together the kaolin slurry and the cement slurry. Initially, a clay slurry was prepared by adding 75% water content to the dry Kaolin (1.5 times Liquid Limit of soil). Then different dosages of cement slurry were added, while the cement slurry was prepared with various mixtures of fresh water and liquid whey waste as presented in Table 1. It should be noted that cement to hydrating liquid ratio of 1 was used. According to Mather and Hime (2002), required time for moisture equalization of clay slurry (i.e. at least 24 h) is much longer than the setting time of cement slurry (i.e. around 30 min), therefore kaolin slurry was mixed one day prior to the final sample preparation. Kaolin powder and fresh water were mixed together for at least 3 min until homogeneous mixture was achieved. The resulting slurry was then stored in air-tight plastic bags for at least 24 h in temperature-controlled room. Once kaolin slurry was ready, preparation of cement slurry was conducted. Cement and fresh water were first mixed for 3 min before adding the predetermined amount of liquid whey waste. The mixing lasted for another 3 min to obtain the designed cement slurry. At the completion of cement slurry preparation, cement and kaolin slurries were mixed together to form the testing mixture. The final mixing lasted around 5 min until color of the testing mixture remained unchanged, indicating homogeneity.

Table 1. Sample preparation

Cement slurry		Clay slurry	Number of prepared samples	
Cement dosage (%)	Whey/Cement (%)	Water/Kaolin (%)	14 day-curing	56 day-curing
10	0	75	3	3
	10			
	20			
	30			
	40			
	50			
	60			
	65			
	70			
	100			
15	0			
	10			
	20			
	30			
	40			
	50			
	60			
	65			
	70			
	100			
20	0			
	10			
	20			
	30			
	40			
	50			
	60			
	65			
	70			
	100			

Specimen was placed in the steel cylindrical moulds with 100 mm height and 50 mm diameter right after the final mixing. Due to high water content of the mixture, soil was placed in the mould using palette knife layer by layer. To reduce damage to the

4   T. Pham Ngoc et al.

specimen during extruding, a thin Petroleum Jelly was applied to the inner wall of the mould, while the base was fully covered by a filter paper. For each specimen, post compaction mass and corresponding density was measured. The density was then used to detect and remove inappropriate specimens that had density clearly different to the average value of the group with the same dosage.

After sample preparation, specimen along with the steel mould were stored in air-tight plastic bags to prevent water evaporation in temperature-controlled room (22 °C) for 24 h to gain enough strength and stiffness for specimen extrusion. Extruded specimens were then cured in the same condition for the designed durations, namely 14 days and 56 days.

2.2 Unconfined Compression Test

An unconfined compression load frame with the loading capacity of 50 kN was utilized to conduct the unconfined compression test. Sample was placed in center of the load cell, covered with cylindrical cap. Axial load was applied at constant loading rate of 1 mm per minute until load values decrease notably with the increasing strain or until 20% axial strain was reached. During the test, readings of stress and displacement were directly transferred to computer for the computation and plotting of stress-strain curves.

Magnitudes of axial stress (σ) and strain (ϵ) in each test were computed from the readings of axial force (F) and axial displacement (h) as below:

$$\varepsilon = \frac{\Delta h}{h_0} \tag{1}$$

$$\sigma = \frac{F}{A} = \frac{F(1 - \varepsilon)}{A_0} \tag{2}$$

where h_0 and A_0 are the initial height and cross-sectional area of the specimen, respectively; A is the average cross-sectional area of the specimen at the corresponding strain, calculated based on the assumption that during the test volume of the specimen remains constant. Once stress-strain curve was obtained, unconfined compressive strength and Young's modulus of each specimen were determined as the ultimate/peak compressive stress and the slope of the initial linear portion of the curve, respectively. Average values were then calculated for different dosages of cement and whey waste. It should be noted that for each dosage, three samples were prepared and tested and the average values are used.

3 Results and Discussion

3.1 Influence of Whey Waste on Strength of Cement Treated Kaolin

Figure 1 presents variations of UCS with increasing portion of liquid whey waste in the cement slurry under three dosages of cement (i.e.10%, 15%, and 20%) and at two curing ages (i.e. 14 days and 56 days). Clearly, curing age had distinct effects on the evolution of strength of cement treated clay for different dosages of cement and whey waste. UCS of 14 day-aged specimens experienced growing reductions with increasing whey waste content (see Fig. 1a). For all cement treated kaolin samples, minor reductions were observed when less than 10% of water in cement slurry was replaced by liquid whey waste, while the most reductions occured when the replacement percentage increased beyond 30%. However, UCS of 56 day-aged specimens experienced notable growth when whey replacement percentage of 10% was adopted, before experiencing consistent reductions at higher replacement percentage (see Fig. 1b).

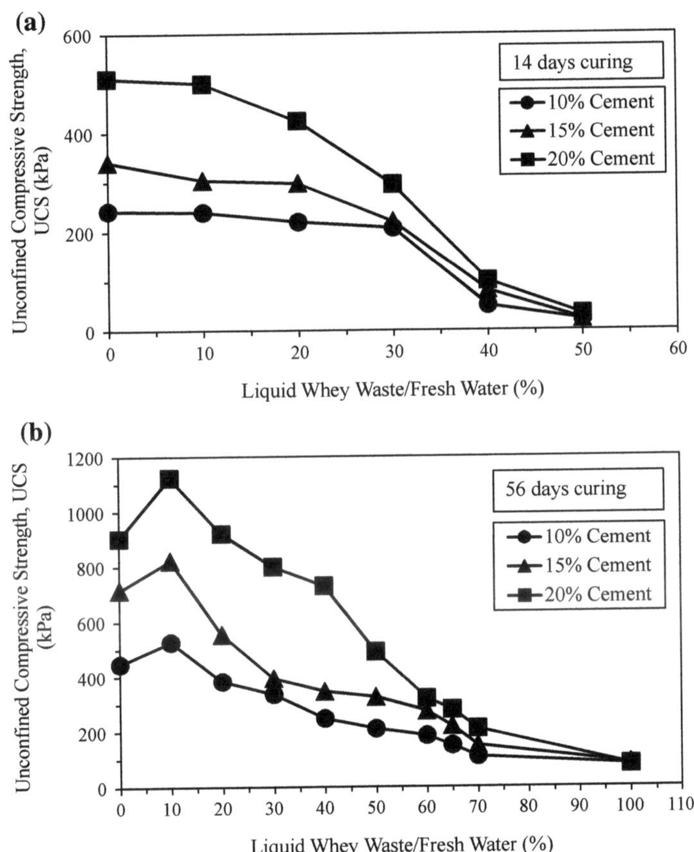

Fig. 1. Effect of whey waste content on UCS of specimens at the curing age of (a) 14 days, (b) 56 days

Time-dependant response of UCS might be influenced by retardation effect of lactose (milk sugar) in liquid whey waste avaiable in the cement paste. Sugar has long been known as a retarding admixture for cement paste (Neville 1995; Taylor 1997). According to Bruere (1966), Thomas and Birchall (1983), and Khan and Baradan (2002), retardation effect of different types of sugar on ordinary Portland cement categorizes into three groups: non-redarding (α-methyl glucoside and α-α trehalose), good retarders (glucose; maltose, celloboise, and lactose), and the most effective retarders (sucrose and raffinose). Based on experimental results, Thomas and Birchall (1983), and Birchall et al. (1984) propsed that the presense of sugar increases the cement solubility, and sugar is then absorbed on to the surfaces of growing hydration products. This then contributes to retarding their growth by allowing much higher concentration of Ca^{2+} ions to coexists in solution with the silicate, hydroxoaluminate and hydroxo-ferrite ions without causing precipitation. It is believed that the retardation effect increases with the amount of sugar, and a large amount of sugar could inhibit the setting of cement paste (Juenger and Jennings 2002; Khan and Baradan 2002; Neville 1995). Interestingly, it was shown in the experiment by Juenger and Jennings (2002) that after some days of delay (e.g. 9 days), degree of hydration of sample treated with 1% sugar increased rapidly, and soon supassed that of the untreated samples. They proposed that once the number of CH and C-S-H nucleation sites becomes larger than the amount of sugar, the "retardation barrier" could be overcome, and therefore newly generated nucleation sites would no longer absorb sugar which is facilating rapid and large evolution of hydration due to high concentration of ions in solution. Alterations in the pore size distribution, increases of surface area of cement paste and decreases in total pore volume due to the presense of sugar, were aslo reported by Juenger and Jennings (2002), suggesting posible effects of sugar on physical and mechanical properties of cement paste. Increases in strength of cement paste due to the presense of sugar was also reported by Neville (1995).

Different responses of UCS at curing age 14 days and 56 day as shown in Fig. 1, can be disclosed by considering the retardation effect induced by the presense of lactose in liquid whey waste. At 14 day-curing age, it seems that the number of generated CH and C-S-H nucleation sites was still smaller than the amount of lactose available in the cement paste. Thus retardation effect had not overcome and would increase with increasing lactose content (i.e. whey waste content). However, at 56 day-curing age, the amount of lactose, contained in the dosage of whey waste less than 10% of cement, was surpassed by the number of generated nucleation sites. Therefore, hydration process was enhanced by the high concentration of ions in the cement paste, resulting in extra evolutions of UCS. At higher dosages of whey waste, the "retardation barrier" would have not broken, thus similar trends in the response of UCS as at 14 day-curing age were observed. Minor differences of USC were observed when water in cement slurry was completely replaced by liquid whey waste suggests extremely low speed of hydration process. The evolution of UCS with time seems to be independent on the cement dosage used.

3.2 Influence of Whey Waste on Stiffness of Cement Treated Kaolin

Figure 2 shows variations of Young's Modulus (E) with the fraction of liquid whey waste replacing water at curing ages of 14 days and 56 days. Similar patterns were observed in comparison with the responses of corresponding UCS (see Figs. 1 and 2). Minor differences, were also observed in the magnitudes of E corresponding to complete replacement of water in cement slurry by whey waste. It is obvious that when the hydration process of cement was too slow, cement content had negligible effects on the stiffness evolution of cement treated clay. At 56 day-curing age, samples with 10% liquid whey waste replacing water in cement slurry gained extra stiffness growths due to the treatment. At this replacement ratio, the effect of high concentration of ions in cement paste would dominate and enhance the hydration process. At higher contents of lactose (i.e. content of whey waste), the stiffness of samples experienced reductions due to the domination of the retardation effect.

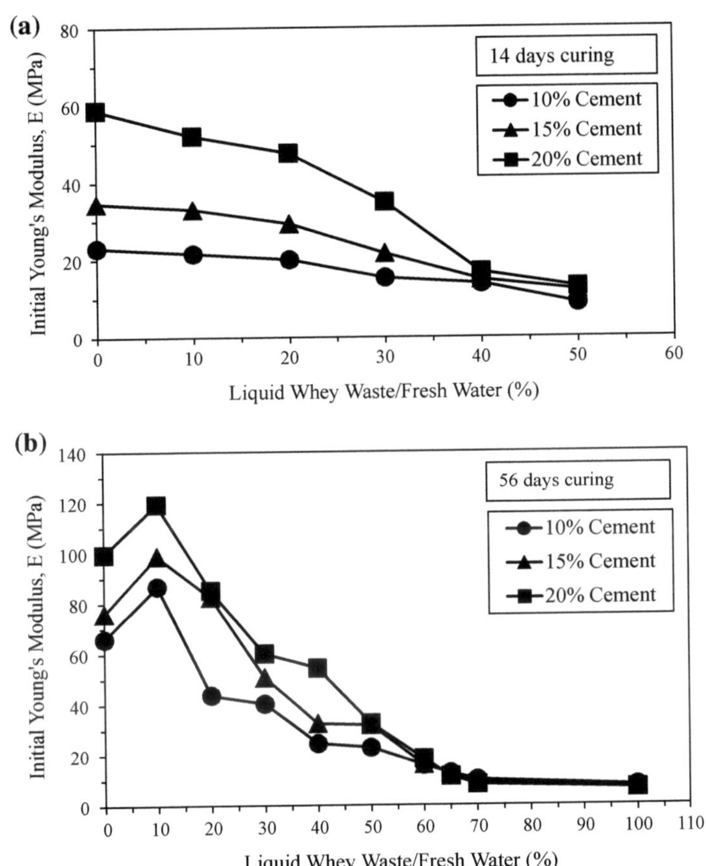

Fig. 2. Effect of whey waste content on E of specimens at the curing age of (a) 14 days, (b) 56 days

3.3 Influence of Cement Content on Strength and Stiffness of Cement Treated Kaolin

According to Figs. 1 and 2, for various whey waste contents, increasing cement dosage caused increase in both soil strength and stiffness. However, the influence of cement content became insignificant when the whey waste replacement approached 100%, corresponding to extremely low speed of hydration process. It is noted that the water/cement ratio of cement slurry of the control sample was 1, thus the percent of whey waste replacing water in cement slurry is equal to the whey waste/cement ratio. Referring to Figs. 1 and 2, it is evident that whey waste/cement ratio plays a key role in capturing the response of cement treated kaolin using liquid whey waste as retarding admixtures.

Figure 3 demonstrates the variation of E/UCS ratio at curing ages of 14 days and 56 days, at the three investigated cement dosages and various liquid whey waste contents. The E/UCS ratio for 14 day-curing age sample was between 75 and 550, with the most variation occurring for the sample treated with 15% cement (Fig. 3a). However, a different picture is presented in Fig. 3b, indicating less variation of the E/UCS ratio (i.e. 66 to 165) at 56 day-curing age with the most significant variation occurred in the sample treated with 10% cement. It is evident that cement dosage actually influences the evolution of strength and stiffness of kaolin treated with cement and liquid whey waste.

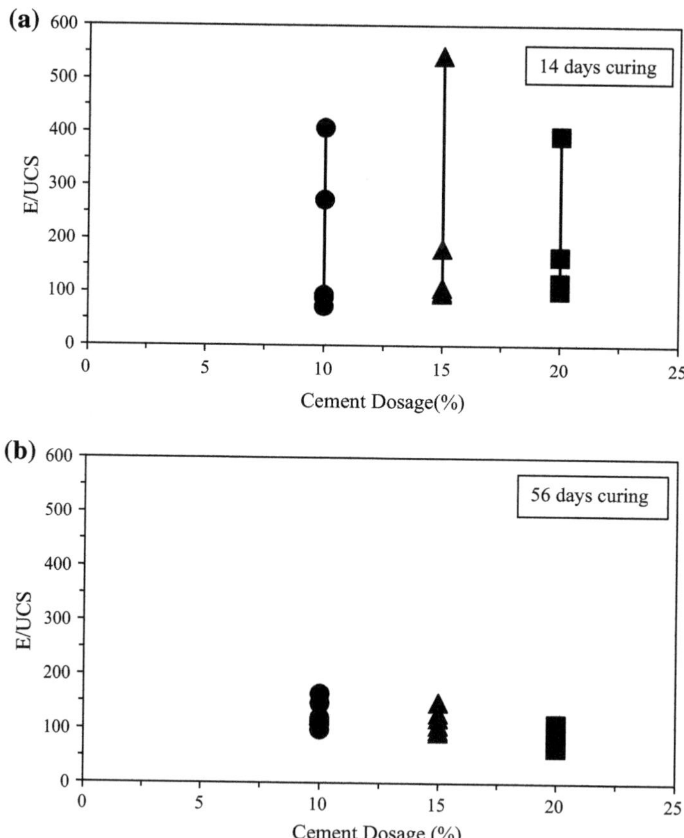

Fig. 3. Variation of E/UCS with cement content at the curing age of (a) 14 days, (b) 56 days, applying various liquid whey waste contents

4   Conclusions

Strength and stiffness of kaolin treated by cement and liquid whey waste at the curing ages of 14 days and 56 days were investigated in a series of unconfined compression tests. The test results confirm the potential application of liquid whey waste by partially replacing water in cement slurry to obtain improvements in workability, strength and stiffness of cemented clay. It was noted that the evolution of strength and stiffness of treated soil depended on curing age, dosages of cement and whey waste content. Adverse effects of the liquid whey addition were observed at 14 day-curing age with decreases of E and UCS when the whey waste content increased. At 56 day-curing age, however, the treatment was effective when 10% of water in cement slurry were replaced by whey waste. The presence of small lactose (milk sugar) in whey waste is believed to result in the observed improvement. Once lactose is added to cement slurry, it is absorbed on the surfaces of generated hydration products, reducing hydration progress, while allows higher concentration of ions coexisting in solution without causing precipitation. Therefore, when the whey content is rather small, the available lactose will soon be exceeded by generated nucleation sites. The retardation effect will then end, and the hydration process will be enhanced owing to the higher concentration of ions in the cement paste. In addition, as the rate of hydration processes are different at before and after the end of retardation effect, curing age could have significant impacts on the variation of strength and stiffness of treated soil.

Acknowledgments. The authors would like to thank Weiqi Li and Willis Susanto for their assistance in conducting the laboratory experiments, as part of their capstone projects at the University of Technology Sydney (UTS). Authors also thank Paesanella (a cheese manufacturer in New South Wales, Australia) for providing us fresh liquid whey waste.

References

Birchall, J., Thomas, N.: The mechanism of retardation of setting of OPC by sugars. Proc. Br. Ceram. Soc. **35**, 305–315 (1984)

Bruere, G.: Set-retarding effects of sugars in Portland cement pastes. Nature **212**(5061), 502–503 (1966)

Cuzman, O.A., Richter, K., Wittig, L., Tiano, P.: Alternative nutrient sources for biotechnological use of Sporosarcina pasteurii. World J. Microbiol. Biotechnol. **31**(6), 897–906 (2015)

Juenger, M.C.G., Jennings, H.M.: New insights into the effects of sugar on the hydration and microstructure of cement pastes. Cem. Concr. Res. **32**(3), 393–399 (2002)

Kelling, K.A., Peterson, A.E.: Using whey on agricultural land–a disposal alternative, Publication-cooperative extension programs, University of Wisconsin Extension (1981)

Khan, B., Baradan, B.: The effect of sugar on setting-time of various types of cements. Q. Sci. Vis. **8**(1), 71–78 (2002)

Lehrsch, G.A., Robbins, C.W., Brown, M.J.: Whey utilization in furrow irrigation: effects on aggregate stability and erosion. Biores. Technol. **99**(17), 8458–8463 (2008)

Marwaha, S., Kennedy, J.: Whey—pollution problem and potential utilization. Int. J. Food Sci. Technol. **23**(4), 323–336 (1988)

Mather, B., Hime, W.G.: Amount of water required for complete hydration of Portland cement. Concr. Int. **24**(6), 56–58 (2002)

McIntosh, G.H., Royle, P.J., Le Leu, R.K., Regester, G.O., Johnson, M.A., Grinsted, R.L., Kenward, R.S., Smithers, G.W.: Whey proteins as functional food ingredients? Int. Dairy J. **8** (5–6), 425–434 (1998)

Modler, H.: The use of whey as animal feed and fertilizer. Int. Dairy Fed. (1987)

Neville, A.M.: Properties of Concrete, vol. 4. Longman, London (1995)

Palmieri, N., Forleo, M.B., Salimei, E.: Environmental impacts of a dairy cheese chain including whey feeding: an Italian case study. J. Clean. Prod. **140**, 881–889 (2017)

Robbins, C.W., Lehrsch, G.A.: Cheese whey as a soil conditioner. In: Handbook of soil conditioners: substances that enhance the physical properties of soil, vol. 1, pp. 167–185 (1998)

Sharratt, W., Peterson, A., Calbert, H.: Effect of whey on soil and plant growth. Agron. J. **54**(4), 359–361 (1962)

Siso, M.G.: The biotechnological utilization of cheese whey: a review. Biores. Technol. **57**(1), 1–11 (1996)

Taube, P., Koglova, N., Orlova, L., Gurevich, E.: Admixture for Mortar and Concrete (1974)

Taylor, H.F.: Cement Chemistry. Thomas Telford (1997)

Thomas, N.L., Birchall, J.: The retarding action of sugars on cement hydration. Cem. Concr. Res. **13**(6), 830–842 (1983)

Experimental Study of Erosion on Expansive Soil Slope Strengthened by HPTRM System

Yingzi Xu[1]($\boxtimes$), Rikui Yan[1], Tang Linqiang[1], Chunyan Kang[1],
and Lin Li[2]

[1] Guangxi Key Laboratory of Disaster Prevention
and Engineering Safety, College of Civil and Architectural Engineering,
Guangxi University, Nanning 530004, China
xuyingzi@gxu.edu.cn, {1052670655, 375405084, 609394677}
@qq.com
[2] Department of Civil and Environmental Engineering,
Jackson State University, Jackson, MS 39217, USA
linli@jsums.edu

Abstract. Rainfall can cause erosion of expansive soil slope and subsequently slope failure. Prior studies have shown that the protection of expansive soil slope against erosion is important. The goal of this study is to investigate the erosion performance of an innovative expansive soil slope strengthening technique – High Performance Turf Reinforcement Mat (HPTRM). The HPTRM is a three-dimensional turf reinforcement mat joined at the intersections of randomly oriented nylon filaments with high tenacity polyester geogrid reinforcement at low strains. As the grass roots grow through the open space of HPTRM, roots become entwined within the turf reinforced mat. The interlocking between roots and HPTRM can enhance the resistance against hydraulic and shear forces during the rainfall-induced erosion. Artificial simulated rainfall system was used to simulate the rainfall intensities and duration in Nanning, China. Different slope gradients and rainfall intensities/duration were studied. The experimental results show that the erosion amount reduced by 70% $\sim$ 85% in the HPTRM-strengthened slope under the moderate-intensity rain or heavy-intensity rainfall, compared to the erosion on the same expansive soil slope without vegetated HPTRM strengthening. The erosion amount was only about 1% of the bare slope when the slope was protected by vegetative HPTRM system under different rainfall intensities. The result demonstrated that that the vegetative HPTRM system can enhance soil resistant ability to rainfall-induced erosion-resistant ability.

1 Introduction

Rainfall can cause erosion of expansive soil slope and slope failure. Expansive soil has strong swell-shrink characteristics, that is easy to crack by the swell-shrink deformation under drying and wetting cycles (Chertkov 2012; Kang 2015). The surface of cracked

© Springer International Publishing AG, part of Springer Nature 2019
W. Frikha et al. (eds.), *New Developments in Soil Characterization and Soil Stability*,
Sustainable Civil Infrastructures, https://doi.org/10.1007/978-3-319-95756-2_2

expansive soil slope can be scoured in rainfall conditions to cause shallow slope instability. Geosynthetics (geocell, geogrid, three-dimensional network geotextile, etc.) can protect erosion of expansive soil slope (Hu et al. 2007; Nong and Deng 2008; Yang et al. 2008; Shi 2009). The flexible protection system combined with geosynthetics and vegetation even can enhance the resistance against the rainfall-induced erosion on soil slope. The flexible protection system combined with geosynthetics and vegetation also can reduce the swell-shrink deformation of expansive soil under drying and wetting cycles (Li 2008; Zhang et al. 2012).

High Performance Turf Reinforcement Mats (HPTRM) is a three-dimensional turf reinforcement mat joined at the intersections of randomly oriented nylon filaments with high tenacity polyester geogrid reinforcement at low strains. As the grass roots grow through the open space of HPTRM, roots become entwined within the turf reinforced mat. The interlocking between roots and HPTRM can enhance the resistance against hydraulic and shear forces during the rainfall-induced erosion. (Kelley and Thompson 2008; Pan et al. 2013; Yuan et al. 2014). Combined with HPTRM and vegetation, the vegetative HPTRM system may reinforce the soil slope to prevent the surface slope failure (Goodrum 2011).

HPTRM system was used in earth dam spillway and the soil slope against erosion by rainfall. The research on HPTRM system was mainly limited in the protective effect of HPTRM system to earth dam. For example, anti-scouring ability of HPTRM system and its ability to protect dam stability were analyzed by full-scale model test and numerical modeling (Hulitt 2012; Xu et al. 2012; Pan et al. 2013; Amini et al. 2013). Based on the full scale flume test and smooth particle hydrodynamics simulation for HPTRM on the levee, relationship between the levee flow velocity and the amount of soil erosion were analyzed (Li et al. 2014, 2015). Their results show that HPTRM is an effective and flexible strengthening material to protect earthen levees against the erosion caused by overtopping flows (Pan et al. 2015a, b; Yuan et al. 2015). However, there is very limited information on the expansive soil slope strengthened by HPTRM system.

In this study, soil erosion physical model tests were conducted to investigate the anti-erosion effect of expansive soil slope strengthened by HPTRM system in different rainfall intensities and duration.

2 Erosion Tests of Expansive Soil Under Rainfall

2.1 Artificial Rainfall System

In this study, the QYJY-501 artificial rainfall system was used to simulate the rainfall intensity and duration for expansive soil slope strengthened by HPTRM system. The rainfall system equipment consisted of main controller, water pipes, water pump, rain shower nozzle, aluminum cylinder barrels of water supply and rain gauge (Fig. 1). The average raindrop size was 0.3–5.7 mm. Simulated rainfall intensity was from 15.0 mm/h to 200 mm/h.

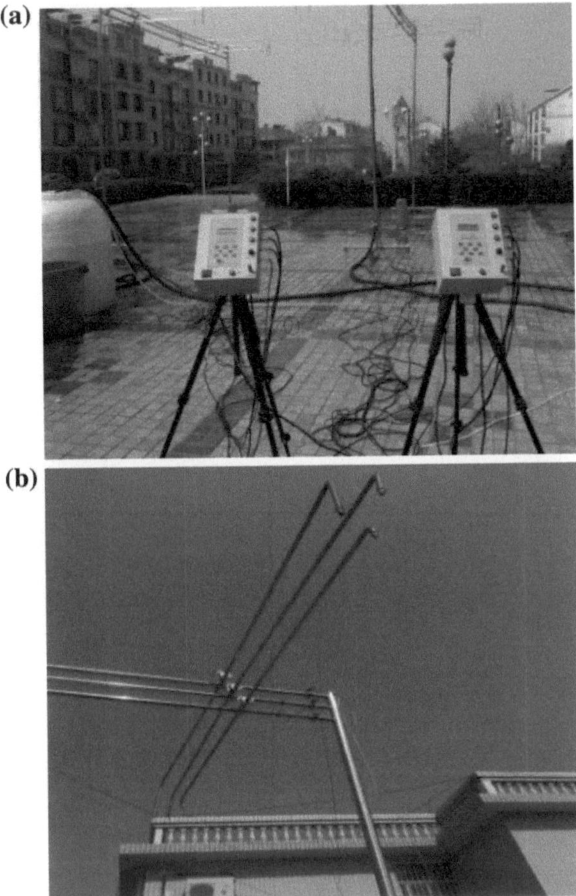

Fig. 1. QYJY-501 Portable artificial rainfall simulator: (a) Controlled panel; and (b) Rainfall simulator.

2.2 Model for HPTRM System Strengthening Expansive Soil Slope

Expansive soil in this study was dark gray clay from Nanning, China. Soil properties are shown in Table 1. The results show that the free swelling rate of the sample soil was 70%, which is in the range of medium-expansive soil. The maximum dry density of soil samples was 1.73 g/cm^3 and the optimum moisture content was 17% based on standard Proctor density test.

Table 1. Soil physical parameter table

Natural density g/cm^3	Dry density g/cm^3	Moisture content %	Liquid limit %	Plastic limits %	Plasticity index	Free swelling ratio %	Permeability cm/s
2.01	1.70	18	48.91	26.46	22.45	70	5.53×10^{-8}

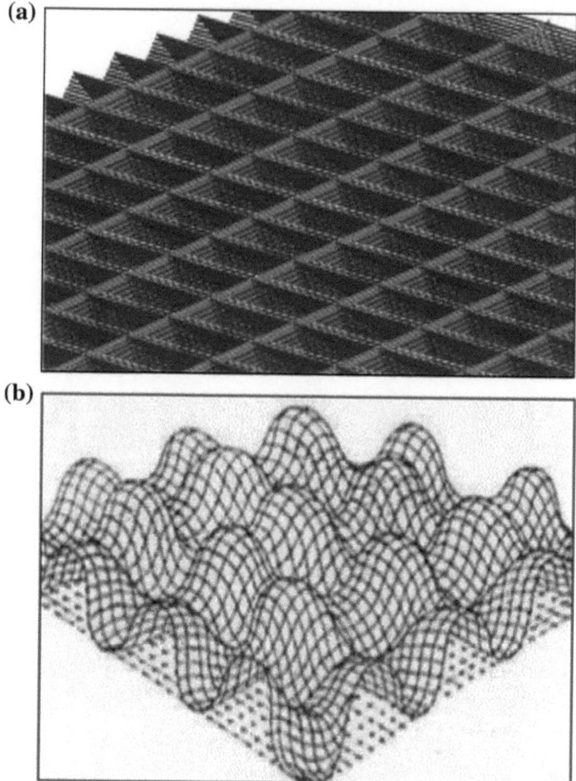

Fig. 2. Geotextile sketch map: (a) HPTRM system, and (b) Three-dimensional network geotextile EM5.

The grass seeds in this test were ryegrass and Bermuda grass, which are often used in soil erosion control programs. These two grasses are suitable for growing in a warm and moist environment like Guangxi. Their germination rate and their root formation were similar (Xu 2014; Kang 2015).

The geotextile material HPTRM was used for this test. The geotextile material EM5 was used as comparison (Fig. 2). The EM5 was a kind of common three-dimensional geotextile mat made from polymer synthetic materials as well as HPTRM, while its radiation resistance, durability and tensile strength were much lower than HPTRM.

The test bed was a rectangular box filled with expansive soil and covered with HPTRM or EM5. The grass seeds mixed with fertilizer and soil were laid on HPTRM or EM5. The geometry of the test bed was 120 cm in length, 60 cm in width, and 30 cm in height. The initial water content of the soil sample was 17%, which is the optimum moisture content. The degree of compaction is 0.9. A drainage pipe was set at the end of the model box to drain rainwater away. The test bed can be changed in different angles to simulate various slope angle. Figure 3 show the detailed information about the test bed and vegetative HPTRM or EM5 reinforced system.

2.3 Experiment Plan

Fig. 3. Images of the HPTRM or EM5 reinforced system: (a) test bed of the system, (b) compacted soil in layer, (c) cover HPTRM or EM5 and sowing seeds, (d) fertilization and (e) grass after 30 d growth.

Rainfall-induced erosion was monitored for the five different scenarios: (1) bare slope; (2) covered with HPTRM; (3) covered with vegetated slope; (4) covered with vegetative HPTRM system; and (5) covered with vegetative EM5 system. In each case, the physical test used the following procedure:

1. Used rainfall intensity of 20 mm/h, 40 mm/h and 60 mm/h under the 20° slope and 60 –120 min of continuous rainfall.
2. Used slope gradients of 15°, 20° and 30° under the rainfall intensity of 40 mm/h and continuous rainfall 60 min.

3 Effects of Rainfall Intensity on Slope Erosion

The total erosion amount of vegetated HPTRM system strengthening expansive soil slope at rainfall intensity of 20 mm/h, 40 mm/h and 60 mm/h was studied. The slope angle was set as 20 degree. The duration of rainfall was 60 min. In comparison, four other expansive soil model with bare soil, soil covered with HPTRM, soil covered with vegetated slope and soil covered with vegetated EM5 were carried out under the same condition. The total erosion amount as a function of rainfall intensity is shown in Fig. 4. The results show that total erosion increased with rainfall intensity at any strengthened condition.

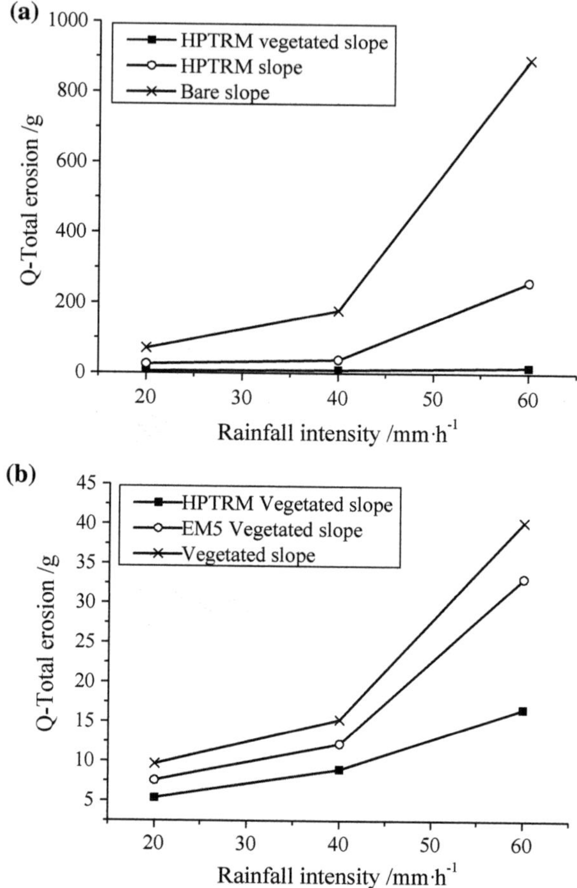

Fig. 4. The relationship between rainfall intensity and total erosion under different slope protections: (a) Different slope protection forms, and (b) Vegetated slope with different geo-materials (the slope was 20°, rainfall duration was 60 min).

As shown in Fig. 4a, the total erosion of soil covered with HPTRM or bare slope increased significantly with the intensity of rainfall increased especially over 40 mm/h. The total erosion of HPTRM vegetated slope was much lower than those of soil covered with HPTRM or bare slope at any rainfall intensity. When the rainfall intensity was less than 40 mm/h, the total erosion of the vegetated HPTRM slope was about 6% of the erosion of bare slope, and was about 25%–35% of erosion of soil covered with

HPTRM. When the rainfall intensity increased to 60 mm/h, the total erosion of the vegetated HPTRM slope was 7% of soil covered with HPTRM slope and was only 1.5% of erosion of bare slopes. Because the expansive soil tends to crack under wetting and drying cycles, there were many cracks in the surface of the bare expansive soil slope in atmospheric environment. The cracked soil is easy to wash away during the rainfall. Therefore, the total erosion of the expansive soil slope without any protection can be 900 g in 60 mm/h rainfall intensity for 20 min. Soil covered with HPTRM can prevent the rainfall-induced erosion to some extent. The interlocking between vegetation roots and HPTRM in HPTRM system can limit cracks of surface expansive soil and enhance the anti-erosion ability. With the increase of rainfall intensity, anti-erosion effect of vegetative HPTRM system strengthened expansive soil slope is obviously better than that of bare or simple HPTRM slope.

Figure 4b show the total erosion of vegetated slope, vegetated EM5 slope, and vegetated HPTRM slope as a function of rainfall intensity. The results show that the total erosion of the vegetated HPTRM slope was less than 50% of erosion of the simple vegetated slope. The total erosion of vegetated EM5 slope was more than that of the vegetated HPTRM slope, but less than that of the simple vegetated slope. When the rainfall intensity increased the erosion of soil covered with vegetated EM5 slope became similar to the erosion of vegetated HPTRM slope. However, the erosion of soil covered with vegetated HPTRM slope in the rainfall intensity of 60 mm/h was less that 15 g, which was only half of the erosion amount of soil covered with vegetated EM5 slope. It shows that the effect of anti-erosion on vegetated HPTRM slope was better than vegetated EM5 slope. The single layer of 3D structural formation of EM5 can be used to explain the weak bonding of grass and EM5 during the rainfall-induced erosion, while the multiple layer of 3D structural formation of HPTRM can provide much strong bonding between the grass root and the HPTRM.

4 Effects of Rainfall Duration on Slope Erosion

The cumulative erosion amount of vegetated HPTRM system strengthening expansive soil slope during 120 min continuous rainfall was studied. The slope angle was set as 20 degree. The rainfall intensity was set at 40 mm/h. In comparison, four other expansive soil model with bare soil, soil covered with HPTRM, soil covered with vegetated slope and soil covered with vegetated EM5 were carried out under the same condition. The cumulative erosion amount as a function of rainfall duration is shown in Fig. 5.

As shown in Fig. 5a, the cumulative erosion amount of slope protected by HPTRM vegetated system was much lower than that of the bare slope or simple HPTRM slope with the duration of rainfall increasing. The cumulative erosion amount of slope protected by HPTRM vegetated system was only about 3% of erosion of the bare slope, or about 25% of erosion of soil covered by simple HPTRM during 120 min continuous rainfall at the rainfall intensity of 40 mm/h. The cumulative erosion amounts of vegetated slope with EM5 and HPTRM are shown in Fig. 5b. The cumulative erosion amount of HPTRM vegetated slope was about 40% of the erosion of simple vegetated

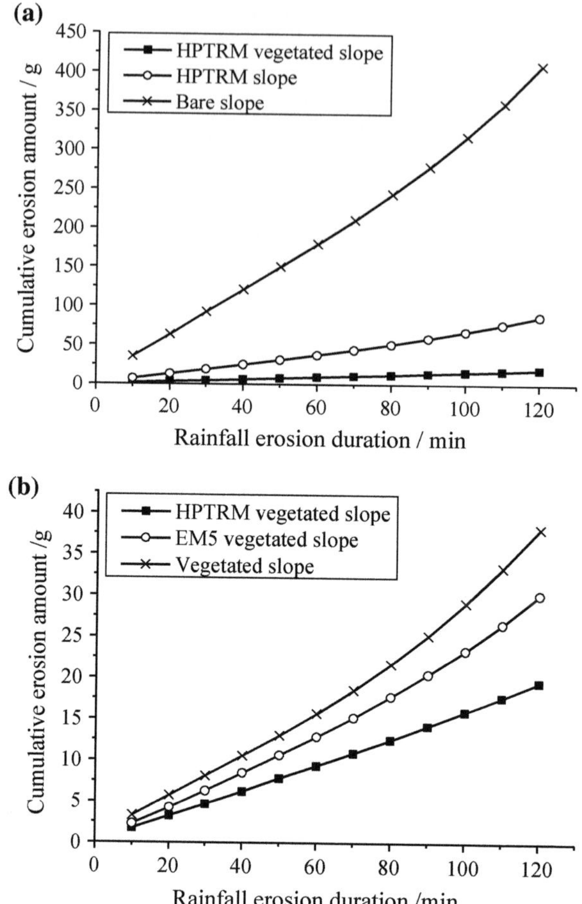

Fig. 5. The relationship between rainfall duration and the cumulative amount of erosion under different slope protections: (a) Different slope protection forms, and (b) Vegetated slope with different geo-materials (the slope was 20°, rainfall intensity was 40 mm/h).

slope and 55% of erosion of EM5 vegetated slope during 120 min continuous rainfall. The results show that anti-erosion effect on vegetated HPTRM protected slope is better than EM5 vegetated or vegetated slope.

5   Effects of Slope Gradient on Slope Erosion

The slope gradient can affect the runoff speed on the surface of slope, thus affect the rainfall-induced erosion. The total erosion amount of vegetated HPTRM system strengthening expansive soil slope at slope gradients of 15°, 20° and 30° was studied. The rainfall intensity was set at 40 mm/h and the duration of rainfall was 60 min. In comparison, four other expansive soil model with bare soil, soil covered with HPTRM, soil covered with vegetated slope and soil covered with vegetated EM5 were carried out under the same condition. The total erosion as a function of slope gradient is shown in Fig. 6.

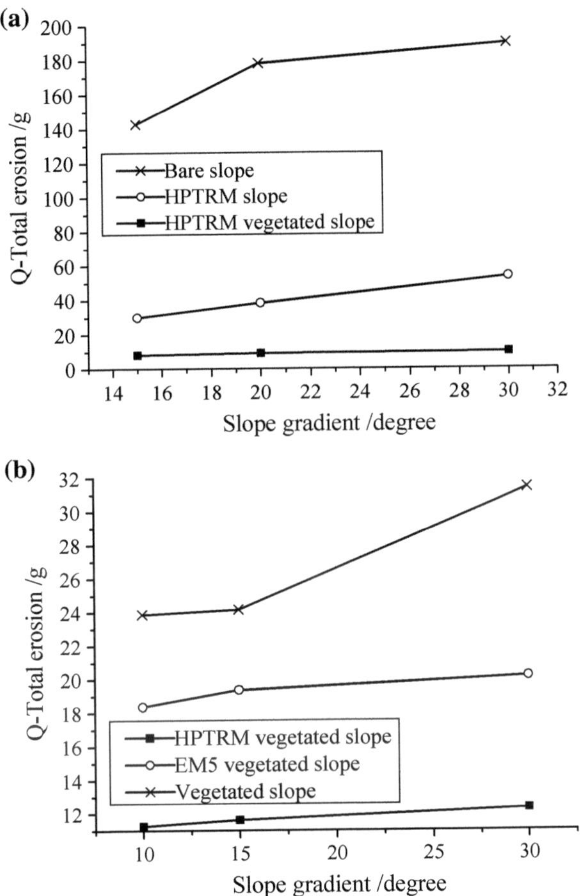

Fig. 6. The relationship between slope gradient and total erosion under different slope protections. (a) Different slope protection forms, and (b) Vegetated slope with different geo-materials (rainfall duration was 60 min, rainfall intensity was 40 mm/h).

As shown in Fig. 6a, the total erosion of soil covered with HPTRM or bare slope increased with slope gradient. The total erosion of HPTRM vegetated slope was much lower than those of soil covered with HPTRM or bare slope at any slope. When the slope increased to 30°, the total erosion of HPTRM vegetated slope was only 10 g, while the total erosion of bare slope was up to 190 g and the total erosion of soil covered with HPTRM was 45 g. Soil covered with HPTRM can prevent the limited rainfall-induced erosion. The interlocking between vegetation roots and HPTRM in HPTRM system can enhance the anti-erosion ability of the slope. With the increase of slope (i.e., the runoff speed of flow), anti-erosion effect of vegetation HPTRM system strengthened expansive soil slope was much better than that of bare or simple HPTRM slope.

Figure 6b show the total erosion of vegetated slope, vegetated EM5 slope, and vegetated HPTRM slope as a function of slope gradient. The results show that the total erosion of the vegetated HPTRM slope was less than 100% of erosion of the simple vegetated slope. The total erosion of vegetated EM5 slope was 50% more than that of the vegetated HPTRM slope, but less than that of the simple vegetated slope. When the slope increased to 30°, the total erosion of HPTRM vegetated slope was only 10 g, while the total erosion of vegetative slope was up to 30 g and the total erosion of soil covered with vegetative EM5 was 20 g. It shows that the effect of anti-erosion on vegetated HPTRM slope was better than vegetated EM5 slope in the various slope gradient. At the steep slope, the increased runoff flow speed and the shear stress had large impact on the vegetative EM5 than the HPTRM vegetated slope, that is related to the relative weak bonding of grass and the EM5.

6   Conclusions

This study used erosion model tests to analyze the anti-erosion effect of expansive soil slope strengthened by HPTRM system subjected to different intensity and duration rainfall. The following conclusions were drawn from this study.

(1) Compared to other forms of protection, vegetative HPTRM system significantly improves the anti-erosion effect on expansive soil slope during all the rainfall conditions. Anti-erosion effect of vegetation HPTRM system strengthened expansive soil slope was obviously better than that of bare or simple HPTRM slope. It is also better than the vegetated EM5 slope and simple vegetated slope. The total erosion of HPTRM vegetated system protection was only 50%–55% of erosion of vegetated slope, about 20%–30% of erosion of HPTRM slope, and only 1%–5% of erosion of bare slope.
(2) Slope erosion increased with higher rainfall intensity and duration of continuous rainfall for slope strengthened with vegetative HPTRM system.
(3) The total erosion amounts of various protection of slope increased slightly with the increasing of slope gradient.

Acknowledgments. This research was supported by National Natural Science Foundation of China (NSFC) (No. 51178124) and the Project of Department of Land and Resources of Guangxi (No. GXZC2015-G3-4366-KLZB).

References

Amini, F., Li, L., Xu, Y.: Slope stability analysis of three earthen levee strengthening systems under hurricane overtopping flow conditions. In: ASCE Geo-Congress (2013). https://doi.org/10.1061/9780784412787.189

Chertkov, V.Y.: Physical modeling of the soil swelling curve vs. the shrinkage curve. Adv. Water Resour. Elsevier (2012). https://doi.org/10.1016/j.advwatres.2012.05.003

Goodrum, R.: A comparison of sustainability for three levee armoring alternatives. In: Koerner, G.R., Koerner, R.M., Ashley, M.V., Hsuan, G.Y., Koerner J.R. (eds.) Optimizing Sustainability Using Geosynthetics. Proceedings of the 24th Annual GRI Conference Proceedings. Dallas, Texas, March 16, pp. 40–47 (2011)

Hu, M.J., Kong, L.W., Guo, A.G., et al.: Expansive soil embankment stability and geogrid treatment effect analysis with strength zoning method. Rock Soil Mech. (2007). https://doi.org/10.16285/j.rsm.2007.09.006

Hulitt, C.D.: Full-scale testing of three innovative levee strengthening systems under overtopping condition. Master thesis, Jackson State University (2012)

Kang, C.Y.: Experimental study of erosion on expansive soil slope strengthened by high performance turf reinforcement mat (HPTRM) system in nanning. Master thesis, Guangxi University (2015)

Kelley, D., Thompson, R.: Comprehensive hurricane levee design: development of the controlled overtopping levee design logic. In: SAME Technology Transfer Conference and Lower Mississippi Regional Conference (Vicksburg, Mississippi), March 2008, pp. 17–19 (2008)

Li, G.W.: Application of geocell slope protection in highway slope of expansive soil area. Highw. Transp. Res. **04**, 46–48 (2008). (Applied Technology Edition)

Li, L., Pan, Y., Amini, F., et al.: Erosion resistance of HPTRM strengthened levee from combined wave and surge overtopping. Geotech. Geol. Eng. (2014). https://doi.org/10.1007/s10706-014-9762-7. Springer

Li, L., Rao, X., Amini, F. et al.: SPH modeling of hydraulics and erosion of HPTRM levee. J. Adv. Res. Ocean Eng. (2015). https://doi.org/10.5574/jaroe.2015.1.1.001

Nong, D.L., Deng, J.Y.: The application of geosynthetics in expansive soil ground treatment. Cities Towns Constr. **6**, 97–98 (2008)

Pan, Y., Li, L., Amini, F., et al.: Full scale HPTRM strengthened levee testing under combined wave and surge overtopping conditions: overtopping hydraulics, shear stress and erosion analysis. J. Coast. Res. (2013). https://doi.org/10.2112/jcoastres-d-12-00010.1

Pan, Y., Amini, F., Li, L.: Failure mechanism of earthen levee strengthened by vegetated hptrm system and design guideline for hurricane overtopping conditions. In: IFCEE 2015 (2015a). https://doi.org/10.1061/9780784479087.227

Pan, Y., Li, L., Amini, F. et al. Overtopping erosion and failure mechanism of earthen levee strengthened by vegetated HPTRM system. Ocean Eng. Elsevier (2015b). https://doi.org/10.1016/j.oceaneng.2014.12.012

Shi, G.L.: Effect of geogrid reinforced expansive soil slope. Master thesis, Wuhan University of Technology (2009)

Xu, S.Y.: Experimental study of expansive soil slope vegetation strengthening system in Nanning. Master thesis, Guangxi University (2014)

Xu, Y.Z., Li, L., Amini, F.: Slope stability analysis of earthen levee strengthened by high performance turf reinforcement mat under hurricane overtopping flow conditions. Geotech. Geol. Eng. (2012). https://doi.org/10.1007/s10706-012-9511-8. Springer

Yang, H.P., Wang, S., He, Y.X.: Treat cut slopes with expansive soils adopting geogrid-reinforced technique. In: Geosynthetics in Civil and Environmental Engineering. Springer Berlin, Heidelberg (2008). https://doi.org/10.1007/978-3-540-69313-0_70

Yuan, S.Y., Li, L., Amini, F., et al.: Turbulence measurement of combined wave and surge overtopping of a full-scale hptrm-strengthened levee. J. Waterw. Port Coast. Ocean Eng. (2014). https://doi.org/10.1061/(asce)ww.1943-5460.0000230

Yuan, S.Y., Tang, H.W., Li, L., et al.: Combined wave and surge overtopping erosion failure model of HPTRM levees: accounting for grass-mat strength. Ocean Eng. (2015). https://doi.org/10.1016/j.oceaneng.2015.09.005. Elsevier
Zhang, Y.J., Wang, G.Y., Wang, L., et al.: Solidify effect of Cut Slope ecology protection with indoor and outdoor tests. J. Chang. Univ. Sci. Technol. **9**(3), 9–14 (2012). (Natural Science)

The Geotechnical Properties, on Water Sensitive Soils, Loess

Gabriela B. Cazacu[✉] and Gabriela Draghici

Faculty of Civil Engineering, Ovidius University, Constanta, Roumania
{brandusacazacu, g.draghici}@yahoo.com

Abstract. This article presents the geotechnical characteristics of loess, water sensitive soil. A comparison of loess from around the world is made (Romania, China). These soils are of quaternary age, are found just below the topsoil and most buildings are founded on them. Problems can arise when the foundation on these lands is softened with water from different sources, permanent or casual. The appropriate parameters of geotechnical solutions for improvement will be presented.

1 Introduction

The article has appeared as a result of the fact that lately there is a globalization of information and workforce. Some international consortia, having activities in different parts of the world, are sending specialists to work on building objectives.

Especially in the field of constructions, a very important problem is the behavior of soils, so comparisons between the same type of soil found in different parts of the world are made.

Loess and loess-like sediments are quaternary soils and cover 10% of Earth's land surfaces.

In Romania, loess and loess-rich soils occupy an area representing about 17% of the territory, and in China the occupied area is 631 000 km^2. In China it is the greatest bulk accumulation of loess on Earth (Liu 1985) and here loess has the greatest thickness.

It is considered that the loess is a soil derived from unstratified wind deposits, the majority of it being comprised of dust. Due to potential collapsibility, loess can be considered as one of the most difficult foundation soils.

Numerous studies and analyses have been done to explain the microstructure, determine the mineralogical composition and geotechnical properties in order to understand the mechanisms leading to its collapsibility.

As there is easy access to information, based on analogies, geotechnical data obtained from different targets in different parts of the world can be used. Although a problematic soil, due to the increased development in the field of construction it is sometimes necessary to build objectives of high importance on loess-rich soils, which are water sensitive.

In order to forecast its behavior when certain initial conditions are changed (accidental damping of the land, earthquakes etc.), knowing and establishing a database of geotechnical parameter values is recommended.

© Springer International Publishing AG, part of Springer Nature 2019
W. Frikha et al. (eds.), *New Developments in Soil Characterization and Soil Stability*,
Sustainable Civil Infrastructures, https://doi.org/10.1007/978-3-319-95756-2_3

Figure 1 shows the location of loess in China (a) Distribution of loess deposits in China and location of the study region of the Central Shandong Mountains. CLP refers to the Chinese Loess Plateau. (b) Chinese Loess Plateau and location of some of the sites mentioned in the text. (c) Location of the studied loess sections in the Central Shandong Mountains (Pingyin — PY; Zibo — ZB; and Qingzhou — QZ) and location of the sediment samples obtained from the lower reaches of the Yellow River (YR). Figure 2 shows the area covered by loess in Romania.

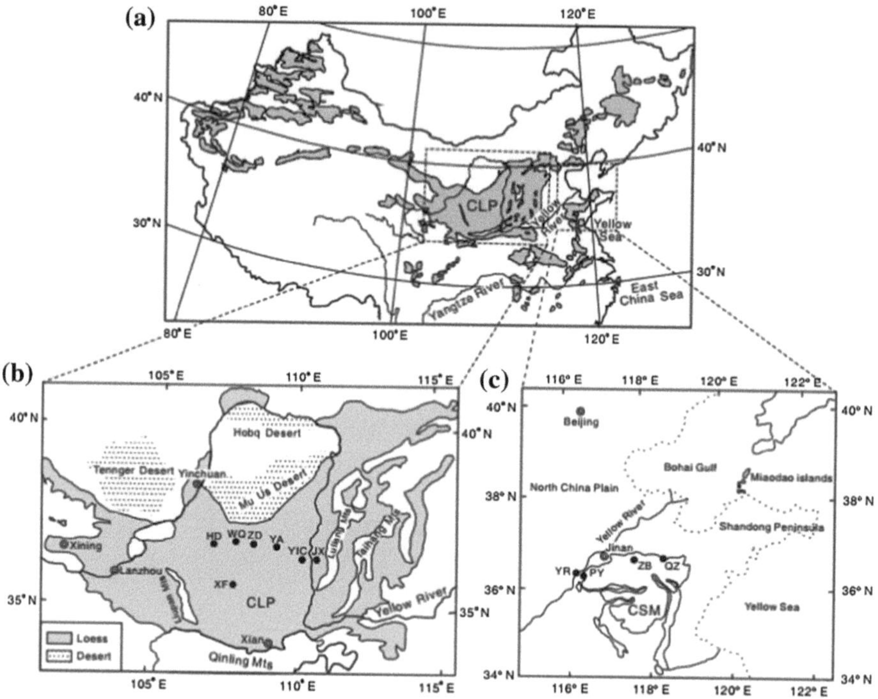

Fig. 1. Loess in China (after Shuzhen et al. 2016)

2   Characteristics of Loess

Loess is a cohesive, high porosity, unsaturated, low humidity soil, which undergoes sudden and irreversible changes of the internal structure when its moisture rises. These changes lead to additional settlements, collapsing, and a decrease in the geotechnical parameter values.

The soils' behavior under the influence of various factors depends on granulometric composition, mineralogical composition and geotechnical parameter values.

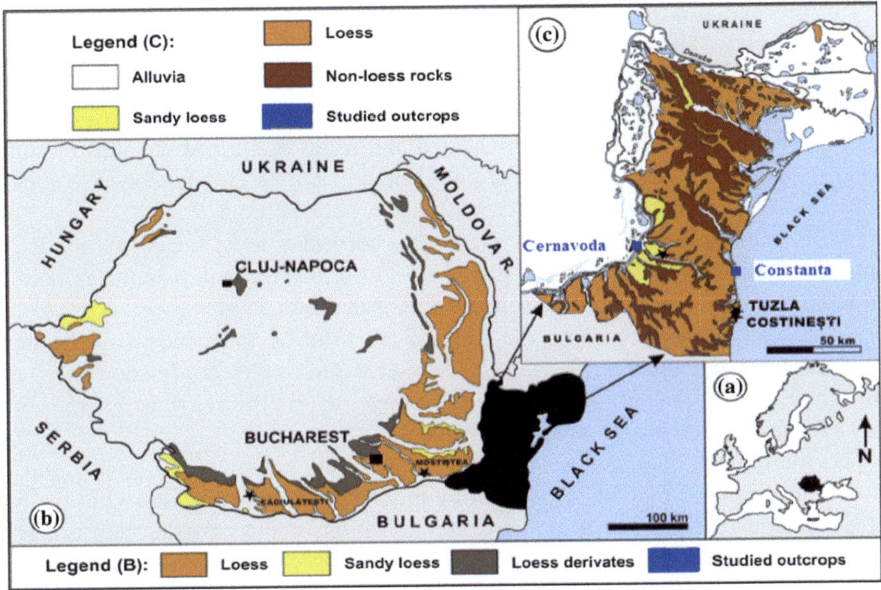

Fig. 2. Loess in Romania (after Constantin et al. 2014)

Table 1. Geotechnical parameters of soils

	Clay <0,005 mm (%)	Silt 0.005 —0,05 mm (%)	Sand >0,005 mm (%)	w_L (%)	w_p (%)	Ip (%)
China	36	54	10	29.53	18.02	11.51
Romania	25	65	10	37	15	22

Although coming from different parts of the world, comparisons can be made between two soils of the same type.

Data presented from remolded Lishi Loess Q2, middle Pleistocene, China and loess of quaternary age obtained from Constanta area, Romania, was used.

Table 1 represents the average geotechnical parameters of the loess.

It is considered that a soil can be customized by considering the granulometric values and plasticity values (which are influenced by granulometry, mineralogical composition, the content of humus). "This can be done using a simple figure called the print".

Definition of the "print" – in order to define the soil nature, the print A applies to Casagrande chart in the 1st dial (point P_1) and a part of the granulometric curve in the 3rd dial; thus for the cohesionless soils we attribute in the granulometric curve the part between the efficient diameter d_{10} (point P_{10}) and d_{90} (point P_{90}), and for the cohesive ones the part between point P_{90} and the 2_μ diameter (point $P_{2\mu}$). In the 2nd dial it may also be represented the integral curve of pores distribution on dimensions both for the soil considered and, eventually, for the geotextile with which it comes in contact.

The 2nd dial is established to represent soil activities, the position of point P_2 indicating, by the slope of the line P_2 O, the activity index $I_A = I_{p/x\ 2\mu}$ defined by Skempton, or the activity fields established by Van de Merwe (1975) under the observation of the behavior of the constructions built in areas with expansive clay.

In the 4th dial the point P_4 is represented and it has a abscissa equal to the upper plasticity limit w_L (%) and the ordinate line related to d_{10} for the non-cohesive soils, or of $d = 2\mu$ for the cohesive ones.

By the union of the points P_1, P_2, P_{90} or $P_{2\mu}$, according to each case, P_4 and P_1, the print A is obtained which form and sizes depend on the common parameters used for characterizing the nature of given soil (Andrei, Athanasiu 1981).

A reference circle is used so the form of the print be independent of the representation scales, the circle having the center in O origin of the coordination axes and a 100 unit diameter, that passes through the points $w_L = 50\%$, $I_p = 50\%$, $X_d = 50\%$, $d = 1$ mm and which has an area of 7854 mm^2.

Considering the fact that any change in the soil nature (solid phase, and the absorption complex), may be interpreted by modifications which occurred in the form and dimension of the print, a global characterization of the material nature may be made to the related area.

A_r = print area/reference circle area

$$A_r = (185 + I_p)\,(W_L + 0, 5_{X2\mu} + 45)/7854 \tag{1}$$

The value of the print for the Chinese loess is 2.27, while the value is 2.47 for the Romanian loess.

An image of the prints obtained for the soils to be analyzed is shown in Fig. 3.

3 Comparison Between the Parameter Values of Shear

Soil behavior is heavily influenced by shearing parameter values, which are used to calculate pressure, the settlement of constructions, the stability of banks etc. Loess-rich soils cause the most problems for the stability of buildings when they are unsaturated and porous.

In This Article, a Parallel Is Drawn Between Soils with Similar Print Values and Mechanical Properties

4 Cohesion

The values for soils with moistures of 10%, 13%, 16% and 18% and dry density values between 1.50 g/cm^3 and 1.70 g/cm^3(3) were considered.

Figure 4 shows the values of cohesion depending on moisture and dry density of loess from both China and Romania (wc – moisture China, wr – moisture Romania).

By all accounts, the cohesion parameter values increase along with the percentage of clay, the increasing dry density and with the decreasing of moisture values.

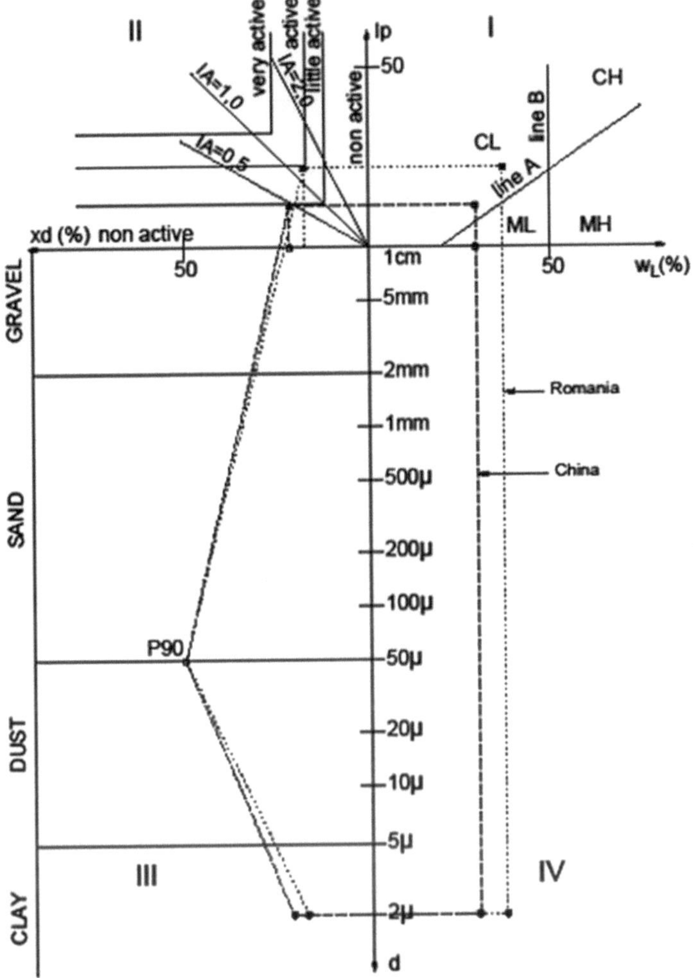

Fig. 3. "Print"soils

Although the soils have approximately equal relative area values, the cohesion values are higher for Romanian soils, although the percentage of clay is higher in Chinese soils.

Cohesion is influenced by the water absorbed by submolecular particles.

It is worth noting that, from a mineralogical point of view, the soils are different, which also results from the fact that the plasticity upper limit of loess in China is lower at a greater amount of clay.

In Romania, for a soil which has in its composition 36% clay and 10% sand, it is defined as dusty clay and the liquid limit values are 47% and plastic limit values 18.6%.

Forecasting the cohesion values and the internal friction angle was tried, using some relations obtained by Chenchen et al. (2017) for China loess.

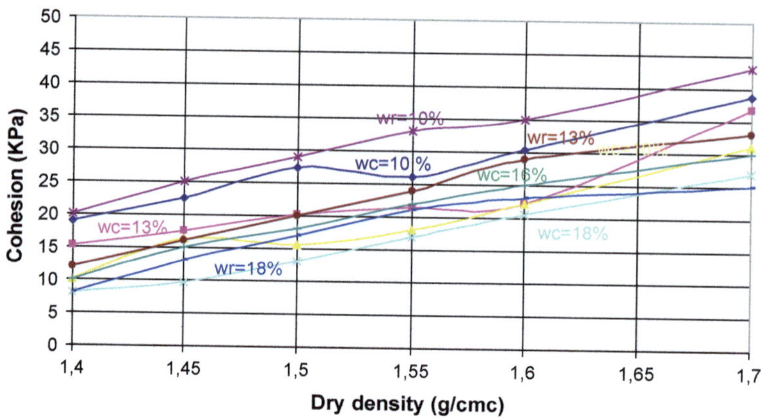

Fig. 4. Variation of cohesion values

Table 2. Relations between the values of cohesion and moisture content

Humidity w (%)	Corelation for cohesion
10	$c = 53.41\rho_d^2 - 102.95\rho_d + 58.98$
13	$c = 228.78\rho_d^2 - 645.72\rho_d + 473.09$
16	$c = 227.81\rho_d^2 - 655.32\rho_d + 486.93$
18	$c = 36.65\rho_d^2 - 39.16\rho_d - 10.55$

The formulas are based on moisture values and the dry density values are taken into account in calculations. They are presented in Table 2.

It is noted that there are differences between −6 kPa and +4 kPa, the large differences are obtained when the loess is dry and stocky, the other values do not differ very much.

For the same loess from China, a relation was found to determine the parameters of resistance, but in calculating it both the values of moisture and the dry density values are taken into account. The formula is shown and denoted by (2).

$$c = 0.035w^2 + 115.921\rho_d^2 + 1.337\rho_d w - 4.361w - 312.217\rho_d + 251.502 \quad (2)$$

Calculating with this relation, the differences obtained between forecasted parameters and actual parameters for loess in Romania are much lower, ranging between ±3 kPa.

Figure 5 shows the obtained differences between the geotechnical parameter values in Romanian loess and the forecasted relationships of Table 2 (Fig. 5a) and the ones forecasted by formula (2) - Fig. 5b.

It is of note that the differences are smaller if the forecast cohesion takes into account the moisture and dry density values.

Large differences, in both cases, are recorded when the moisture is low, w = 10%, and the dry density is around 1.55 g/cm^3.

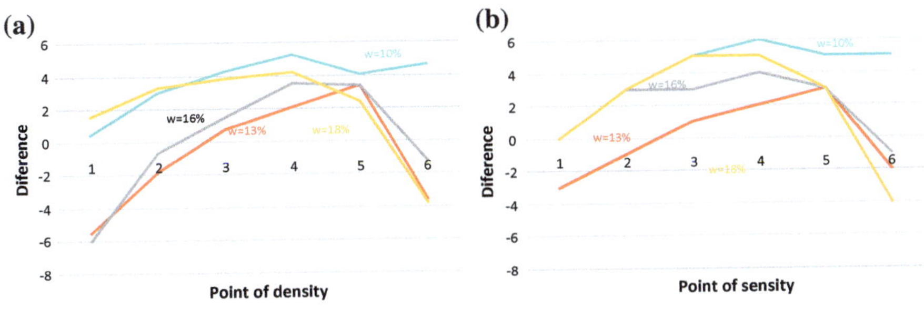

Fig. 5. XXX

It is confirmed that the cohesion parameter values increase along with the percentage of clay, the dry density and with the decrease of moisture values.

4.1   Internal Friction Angle

For the analyzed loess-rich soils, friction angle values based on the moisture and density were obtained, and are shown graphically in Fig. 6.

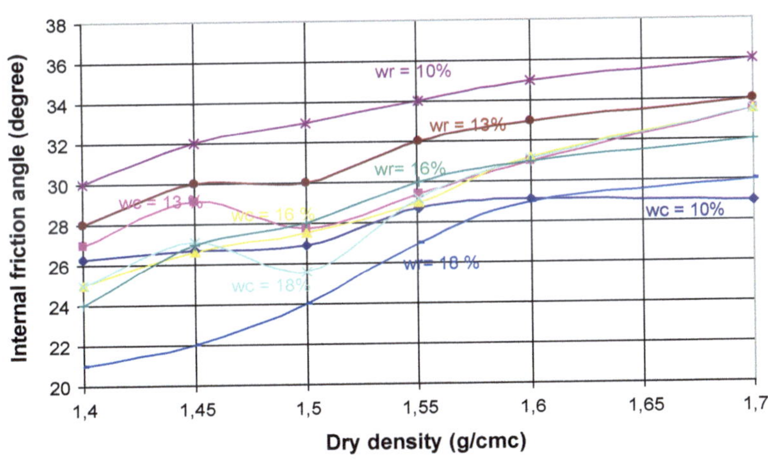

Fig. 6. Variation of internal friction angle value

It is found that for the loess in Romania the angle of internal friction values are between 21% and 36% and between 25% and 34% for China.

For the analyzed loess from China, the friction angle values are lower, due to higher clay percentage from the granulometric composition.

If the same value of density of the soil in the dry state is considered, having 10% moisture values, the angle of friction values for loess in Romania are about 30 degrees and an increase of the moisture lowers the values to 21 degrees.

An attempt was made to verify the relations obtained by Chenchen Huo et al. and are presented in Table 3. The values of the angle of internal friction will only be determined by the moisture of the soils.

Table 3. Relations between the values of the angle of internal friction and dry density

Density ρ (g/cm^3)	Corelation for cohesion
1.5	$\theta = -0.10w^2 + 2.68w + 10.1$
1.6	$\theta = -0.05w^2 + 1.65w + 17.6$
1.7	$\theta = -0.25w^2 + 7.25w - 18.5$

For the Chinese loess analyzed, a correlation to determine the angle of internal friction value has been determined, to which the moisture and dry density values are analyzed, formula 3.

$$\theta = -0.093w^2 + 10.997\rho_d^2 + 2.303\rho_d w - 0.878w - 45.597\rho_d + 54.601 \qquad (3)$$

Although there are differences in terms of the variation of the angle of internal friction according to the moisture and dry density values, the relations used have resulted in large variations of up to 7 degrees, only on the low moisture soils of Romania, otherwise variations were up to ± 2 degrees.

5   Conclusions

Two soils, with relatively close print values, although from different places on Earth, may have similar mechanical properties.

When no data on geotechnical parameter values exists: either not available, undisturbed samples could not be obtained or some were damaged, existing analogies between soils can be used.

Loess-rich soils are water sensitive and can sometimes have settlements of about 10 cm/m in the case of its moistening without being burdened by a building.

Therefore in the design stage it is important to foresee their behavior under changes to the initial conditions.

In addition to the "in situ" data obtained for objectives achieved in certain areas, data from specialty literature from other areas of the world may be used.

References

Cazacu, G.B., et al.: The systematization and storing methods of information concerning the geotechnical parameters—XIII. In: Danube—European Conference on Geotechnical Engineering, Ljubljana, vol. I, pp. 7–11 (2006)

Constantin, D., et al.: High resolution OSL dating of the Costinesti section (Dobrogea SE Romania) using fine and coarse-quartz. Quat. Int. (2014)

Derbyshire, E.: Recent research on loess and palaeosols, pure and applied: a preface. Earth-Sci. Rev. **56**, 1–4 (2001a)

Derbyshire, E.: Geological hazards in loess terrain, with particular reference to the loess regions of China. Earth-Sci. Rev. **56**, 231–260 (2001b)

Chenchen, H., et al.: Experimental study on shear strength indices of unsaturated remolded Lishi loess–Yellow River, vol. 39, no. 3, pp. 137–141 (2017)

Shuzhen, P., Qingzhen, H., Luo, W., Min, D., et al.: Geochemical and grain-size evidence for the provenance of loess deposits in the central Shandong mountains region, Northern China. Quat. Res. **85**, 290–298 (2016)

Wu, C., Ye, G., Zhang, L.: Depositional environment and geotechnical properties of Shanghai clay. Bull. Eng. Geol. Environ. **74**(3), 717–732 (2015)

Ling X.: Landslides in a Loess Platform, North-west China (2014)

Experimental Study on Wet and Dry Cycle Durability of Solidified Silt by Zeolite and Cement

Dongmei Zhang[1(✉)], Baotian Wang[1], and Cheng Feng[2]

[1] Key Laboratory of Ministry of Education for Geomechanics and Embankment Engineering, Hohai University, Nanjing, China
zhangdongmei_hhu@163.com, btwang@hhu.edu.cn
[2] Civil Air Defense Institute of Kingdom Architectural Design, Nanjing, China

Abstract. This paper presented a study on wet and dry durability of solidified silt, by using zeolite and cement for solidification. The wet and dry test was conducted to measure the changes of mass, volume and unconfined compression strength (UCS) with different zeolite and cement content. Furthermore, the influence of zeolite and cement on the physical and mechanical properties of solidified silt will be discussed. The results show that cement plays an important role in improving the wet and dry durability of solidified silt. The volume of solidified silt is slightly expansible with the addition of zeolite at low cement content, but the expandability can be substantially offset as the cement content increase to 20%. With the change of cement content, the addition of zeolite has different effects on the wet and dry cycle durability of solidified silt: zeolite can weaken the wet and dry cycle durability of solidified body when the cement content is less than 20%, while zeolite can enhance the wet and dry cycle durability of solidified body when the cement content is more than 20%.

Keywords: Zeolite · Cement · Durability · Silt solidification
Wet and dry cycle

1 Introduction

Solidification is known as one of the most important techniques for the treatment of silt. The solidification technology of silt mainly includes lime curing, large encapsulation, special agent curing and cement curing [1]. Among these technologies, cement curing is the mostly widely used, due to relatively low cost and better curing effect.

With the increasing of industrial sewage discharge, the content of heavy metals in sediment of urban rivers is getting increased [2]. When cement is used to solidify silt, the crystals formed by the hydration reaction can block heavy metals in solidified body [3, 4]. But a large amount of cement will be needed for solidification if cement is used as the only curing agent. Besides, fixed heavy metals will be susceptible to re-leaching due to rainfall, which will inevitably bring the increase of cost and secondary pollution [5, 6]. Therefore, It is necessary to add admixture with cement in order to reduce the cost and get heavy metal effectively fixed. A number of studies have been focused on the use of cement-based stabilization/solidification for the treatment of silt contamination by heavy metals [7–9], which enable sludge recycle and reduce the hazard caused by heavy metal leaching.

Zeolite is composed of hydrous aluminum silicates, which can be effectively employed as adsorbent for many wastewater pollutants [10]. Especially, they present a special adsorption of heavy metal ions due to the fact that they exhibit a lot of pore with uniform size and enormous surface area [11]. Several researchers have proved the significant stabilizing effect of using zeolite for the treatment of heavy metals in silt [12]. Therefore, this paper proposed the use of natural clinoptilolite combined with cement for the resource processing of silt contamination by heavy metals, in order to achieve high strength and better heavy metals stability at lower cement content.

At present, there are few studies on the long-term stability of solidified silt in China. Li Lei et al studied the cement-solidification of sludge using bentonite as additive. The wet and dry cycle test was conducted to measure the parameters of relative mass loss, volume change and unconfined compression strength (UCS) under the condition of violent water changes, in order to evaluate the long-term stability of the solidified sludge [1]. The durability of cured silt is susceptible to the wet and dry conditions. And the physical-mechanical properties under wet and dry cycle are the important indexes to evaluate the long-term stability of solidified silt.

Based on these, it is meaningful to study physical and mechanical properties of solidified silt under the wet and dry cycle test firstly. In addition, the long-term stability of the cured sludge can be the further research. The aim of this study is to clarify the influences of different zeolite and cement content on the wet and dry durability of solidified silt, according to the quality changes, volume changes and unconfined compression strength changes of solidified silt in the process of wet and dry cycle test. At the same time, this study will provide the basic basis for the application of solidified silt using zeolite and cement.

2　Materials and Methods

2.1　Materials

The silt used for the experiments originates from Wuxi Taihu Lake, which is placed in a plastic bag to retain its initial moisture content until the start of test. The main basic properties of silt are shown in Table 1. The curing agent used in this work is 32.5# Yuhua ordinary Portland cement. The natural clinoptilolite which is used for addition agent originates from Shandong Jiaozhou. Its composition and content are shown in Table 2.

Table 1 Basic properties of silt

Moisture content/%	Liquid limit/%	Plastic limit/%	Plastic index/%	Organic matter content/%
80	45.0	23.0	22.0	4.6
Soil classification/%	Heavy metal content/(mg/kg)			
CH	Cu	Pb	Zn	Cr
	80	38	60	60

Table 2 Composition and content of Zeolite

Composition	SiO_2	Al_2O_3	MgO	Fe_2O_3	K_2O	Na_2O
Content/%	69	12.49	1.23	1.03	2.87	0.49
Composition	MnO_2	TiO_2	FeO	P_2O_5	F	CaO
Content/%	0.03	0.31	0.19	0.02	0.02	3.07

2.2   Method

All the samples were prepared by mixing zeolite and silt, then adding cement. The ratio between them was calculated according to the previous work. Based on Zhong Xuecai and He Jianmin [9, 13], 10%, 15%, 20% and 25% of cement were proper for solidification. At the same time, according to the study of Wang Feng and Xie Hualin [14–17], it is estimated that about 6% zeolite is appropriate to fix heavy metals in sludge. Therefore, the zeolite content is 0%, 4%, 6%, 8%. The ratio of curing agent is presented in Table 3.

Table 3 Ratio of curing agent content with the different content of zeolite

Ratio	1				2			
Cement content/%	10	15	20	25	10	15	20	25
Zeolite content/%	0	0	0	0	4	4	4	4
	3				4			
Cement content/%	10	15	20	25	10	15	20	25
Zeolite content/%	6	6	6	6	8	8	8	8

Based on the above ratio, the prepared mixtures were put into the mold with diameter 39.1 mm and high 80 mm. Four parallel samples were prepared for each ratio. There are 64 experiments totally. All samples were cured for 28 days. Before the wet and dry test, the unconfined compressive strength, moisture content, and mass were measured. Then, 12 times wet and dry cycle were carried out. The surface cracking, mass loss, volume and unconfined compressive strength of the solidified body were observed and measured after every cycle.

The test indexes include: mass percentage M, volume percentage S and unconfined strength decay rate R, which reflect the mass changes, volume changes and strength changes of cured silt respectively. And the softening coefficient λ is used to compare the water stability of cured silt after 12 times of wet and dry cycle. The effect of cement and zeolite content on the water stability of cured body will be further revealed. The test indexes are defined as follows:

$$M = \frac{W_i}{W} \times 100\% \tag{1}$$

$$S = \frac{V_i}{V} \times 100\% \tag{2}$$

$$R = \frac{q_{ui}}{q_u} \times 100\% \tag{3}$$

$$\lambda = \frac{q_{u12}}{q_u} \times 100\% \tag{4}$$

W_i, V_i and q_{ui} represent the mass, volume, and unconfined compressive strength of cured body after the i th wet and dry cycle respectively. W, V and q_u represent the mass, volume, and unconfined compressive strength of the cured body before the wet and dry cycle respectively. q_{u12} represent unconfined compressive strength of the cured body after 12 times wet and dry cycle. When 12 times wet and dry cycle are carried out, the value of R is equal to the value of λ.

3   Results and Discussion

3.1   Adsorption of Heavy Metals in Solidified Silt by Zeolite and Cement

Before starting the wet-dry cycle test, this study also followed the Chinese toxic leaching test method GB5086.2-1997 to do the short-term oscillation test, in order to study the adsorption of heavy metals by zeolite and cement in solidified silt. the cement content is 20%; the zeolite content is 0%, 4%, 6%, 8% respectively. As shown in Fig. 1, the results present that the incorporation of zeolite effectively reduces the precipitation of total heavy metals in solidified silt; and improves the stability of heavy metals under short-term immersion. Then, the influence of zeolite and cement on wet and dry durability of solidified silt is investigated.

3.2   Mass Changes of Cured Silt During Wet and Dry Cycle

The mass changes of cured silt during wet and dry cycle experiment are shown in Fig. 2. From Fig. 2(a) it can be noticed that when cement content is 10%, the mass percentage of cured body decreases to 80% as the number of wet and dry cycle increases. In this case, the addition of zeolite aggravated the mass percentage loss of cured body as a whole: The higher the amount of zeolite, the faster the mass percentage of cured body decreased after 8 times of wet and dry cycle; the mass percentage of

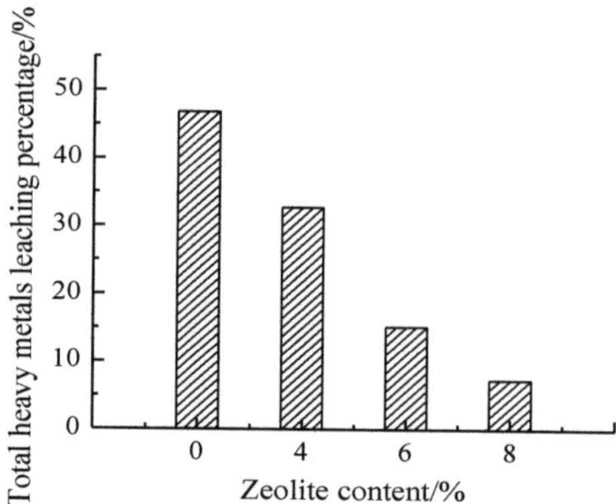

Fig. 1 Total heavy metals leaching percentage of solidified silt with the different content of zeolite

cured body decreased about 50% after 12 times wet and dry cycle. This phenomenon can be interpreted by the following reasons: during the drying process, a large amount of hole water in zeolite is evaporated, and when it is immersed in water again, the zeolite with enormous surface area absorbs water quickly. Since water enters into the cured body rapidly, the compact structure between particles is washed. The surface erosion is exacerbated due to the decrease of the cohesive force between the particles.

In comparison to Fig. 2(a), the mass loss of cured body is significantly reduced due to the increase of cement content. As shown in Fig. 2(b), the maximum mass loss decreased about 15%. The increase of cement content improved the wet and dry cycle durability of cured body. When the cement content exceeds 20%, the mass percentage change is controlled within 5%. As the cement content is increased to 25%, the mass loss of cured body adding 8% zeolite is less than that of single cement. In this case, the cured body can still remain intact after 12 times wet and dry cycle, which indicate that the addition of zeolite can strengthen the cured body to resist the damage caused by wet and dry cycle.

3.3 Volume Changes of Solidified Silt During Wet and Dry Cycle

The volume change of cured silt, as shown in Fig. 3, indicate the following:

(a) after 12 times of wet and dry cycle, the volume loss is 11% when mixing with cement only, while the volume loss is 33% when adding zeolite. This result show that the addition of zeolite increases the volume loss of cured body when the cement content is 10%;

(b) when the cement content is 15%, the maximum volume change is controlled within 5%, which show that the increase of cement content effectively improve the ability to resist the damage caused by wet and dry cycle. When the zeolite content increase from 6% to 8%, the cured body reflects the trend of expansion and the volume increased by nearly 6%.

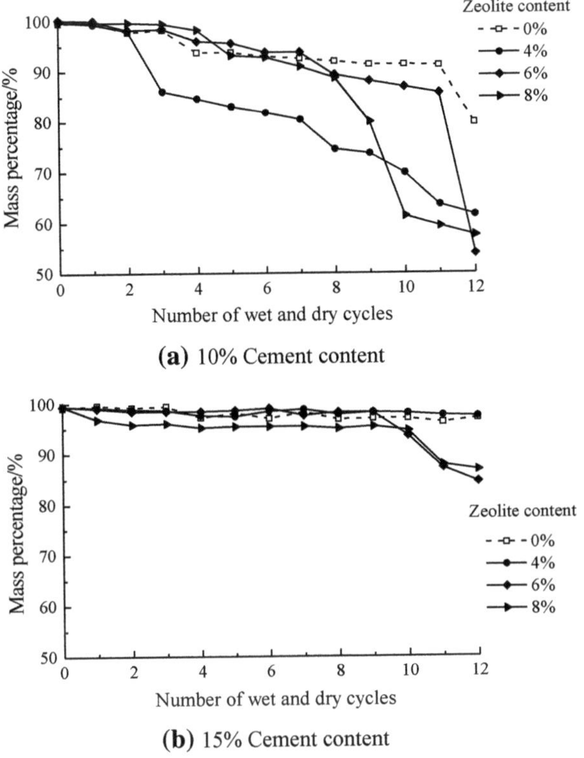

(a) 10% Cement content

(b) 15% Cement content

Fig. 2 Mass changes of cured silt during wet and dry cycle

The above phenomenon is mainly due to the addition of zeolite with expansibility [18]. In comparison to 15% cement, the particles can't be effectively cemented when 10% cement, and the solidified body is eroded during wet and dry cycle. In this case, the expansibility of zeolite is not reflected. Therefore, the expansibility of cured body can be controlled by the amount of cement.

When the cement content is 20% and 25%, the volume loss is relatively reduced, remaining at 4% basically. And the more the cement content, the smaller the volume loss of solidified body during wet and dry cycle. From the test results it can be obtained that solidified particles have a strong cohesive force to effectively offset the swelling effect by zeolite when the cement content exceed 20%. In this case, the volume and mass loss of solidified body is small in wet and dry cycle, which can meet the requirement of durability of cured silt for project.

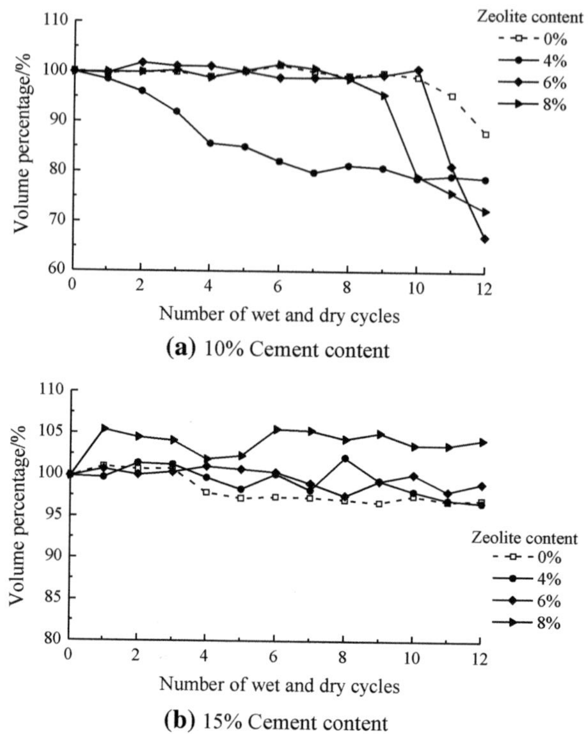

(a) 10% Cement content

(b) 15% Cement content

Fig. 3 Volume change of solidified silt during wet and dry cycle

3.4 Unconfined Compression Strength Changes of Solidified Silt During Wet and Dry Cycle

The relationship between the unconfined strength of solidified body and the increase of zeolite content is given in Fig. 4. The unconfined compressive strength, as shown in Fig. 4(a), was measured before the first wet and dry cycle. When the cement content is 25%, the strength increases to 178 kPa with the increase of zeolite content. (According to the engineering indicators, when the unconfined compressive strength of curing soil is greater than 100 kPa, it will meet the general engineering requirements of filling) [19].

It can be seen from the Fig. 4(b): the strength decreases with the increase of zeolite content from 6% to 8% when the cement content is less than 15%. On the contrary, when the cement content $\geq 20\%$, the strength showed a tendency to increase with the increase of zeolite content as shown in Fig. 4(c) and (d). The law of strength change measured after 8 times and 12 times of wet and dry cycle is consistent.

The main conclusion that can be drawn from all the above is as follows: (1) the effect of zeolite on the strength of solidified body is two-sided and 20% cement content is the critical point. (2) This property emerge gradually with the increase of the number of wet and dry cycle, which did not appear before the test. The main reasons are as

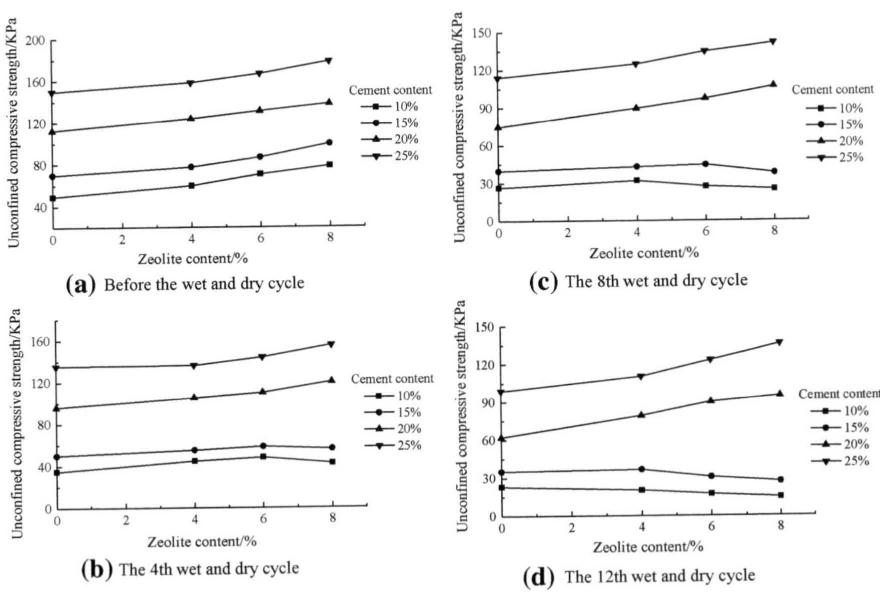

(a) Before the wet and dry cycle   **(c)** The 8th wet and dry cycle

(b) The 4th wet and dry cycle   **(d)** The 12th wet and dry cycle

Fig. 4 Unconfined compression strength changes of solidified silt with different content of zeolite

follows: with the increase of cement content and the deepening of hydration reaction, the elevated $Ca(OH)_2$ concentration makes the added zeolite spin off the silicon. Furthermore, its active ingredient reacts with $Ca(OH)_2$ to produce CAH and CSH, which can further enhanced the strength. When the cement content reaches 20%, the space grid structure caused by hydration reaction become more secure and compact due to the filling effect of zeolite on the pores of solidified silt. In this case, the unconfined compressive strength will increase.

Since cement has a major effect on the unconfined compressive strength of cured body, it is necessary to study the effect of cement content on strength under wet and dry cycle. Figure 5 shows the change of strength decay rate R with different cement content, when zeolite content is 8%. As shown in Fig. 5, the strength decay rate R decreases rapidly (with the number of wet and dry cycle) when cement content is 10% and 15%. In contrast, when the cement content increases to 20%, the decline of strength decay rate significantly slow down and the wet and dry durability of solidified silt significantly enhance. Furthermore, the law of strength decay rate of 25% cement content is similar to that of 20% cement content. The results of previous studies [20] also show that the curing effect of low cement content is not enough. And until 20% cement content, the solidified body will have sufficient strength. The results of this study are also consistent with this.

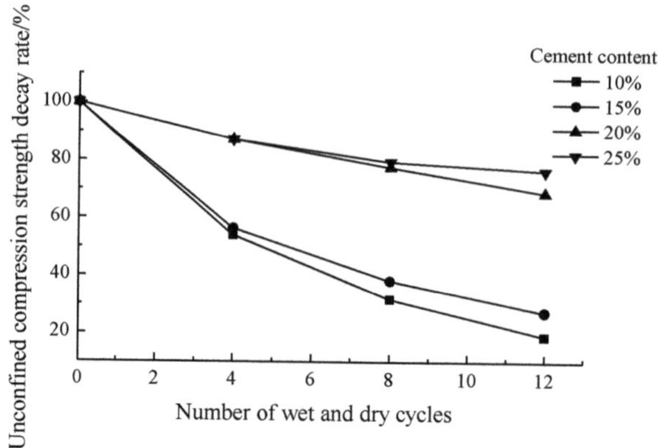

Fig. 5 Unconfined compression strength decay rate of solidified silt during wet and dry cycles

In summary, the cured body can resist the effect of alternating wet and dry cycle until 20% cement content and at this time the addition of zeolite can improve its durability.

Softening coefficient is one of the important indexes to determine the stability of rock water [21]. This paper intends to define the softening coefficient to comprehensively reflect the stability of solidified silt. The relationship between the softening coefficient of cured body and the zeolite content is shown in Fig. 6. With the increase of zeolite content, the softening coefficient of cured body show the following polarization trend: When the cement content does not exceed 15%, the softening coefficient decreases with the increase of zeolite content. However, when the cement content exceeds 15%, the softening coefficient increases with the increase of zeolite content. Based on these results, it can be interpreted that when the cement content is small, the zeolite weakens the water stability of solidified body. But when the cement content is enough, the zeolite has a significant beneficial effect on the stability of solidified body.

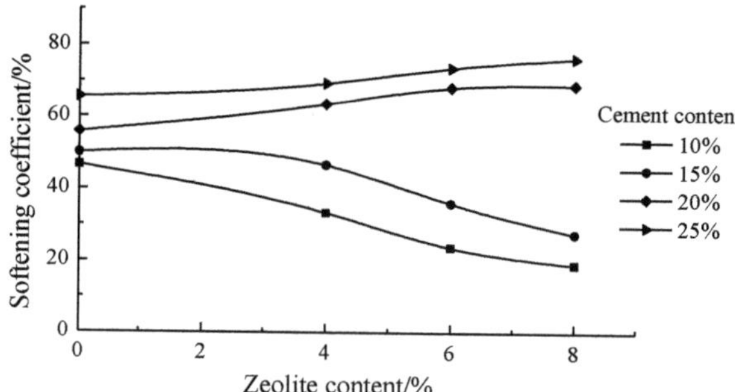

Fig. 6 Softening coefficient change of solidified silt with the different content of zeolite

4 Conclusions

The addition of cement can significantly improve the unconfined compressive strength of solidified body, and reduce the loss of mass and volume of the solidified body in wet and dry cycle. The cement plays the major role in improving the wet and dry durability of solidified silt.

When the cement content is 15% and zeolite content is increased to 8%, the volume of solidified body increases by nearly 6%. When the cement content continued to increase to 20%, the increase of cement content offset the expansion caused by zeolite. Thus, the expansion can be controlled by incorporating a sufficient amount of cement.

When the cement content exceeds 20%, the addition of zeolite can effectively improve the durability. The unconfined compressive strength and water stability of cured body are enhanced with the increase of zeolite content.

Conclusively, it can be suggested that the treatment of zeolite and 20% cement is sufficient to solidify silt and to produce a theoretical basis for the long-term stability of sludge solidified by zeolite and cement.

The work in this paper is only limited to the study of wet and dry cycling in long-term stability. The research on the long-term stability evaluation indexes such as freeze-thaw cycle and heavy metal ions stability need to be further studied.

References

1. Lei, L.: Study on sludge solidification technology and heavy metal pollution contorl. Hohai University (2006)
2. Wei, Z., Fan-Lu, M., Yi-Yan, L., Sheng-Wei, W., Zheng, S., Chun-Lei, Z., Lei, L.: Subject of "mud science and application technology "and its research progress. Rock Soil Mech. (11), 3041–3054 (2013)
3. Yang, Q., Wei, Z., Jian-Ping, B., Qing-Song, L., Jian-Ming, C.: Application of polluted sediment solidification/stabilization technology in Ping lake of Hengyang City. Environ. Sci. Technol. **34**(6), 137–140 (2011)
4. Gougar, M.L.D., Scheetz, B.E., Roy, D.M.: Ettringite and C-S-H Portland cement phases for waste ion immobilization: a review. Waste Manag. **16**(4), 295–303 (1996)
5. Huaming, X., Guangming, Z., Wenlian, L., Chuan, W., Jing, H., Haiyin, X., Zhaohui, Y.: Stabilization/solidification of heavy metal contaminated sediment using cement, fly ash and DTCR. Chin. J. Environ. Eng. **7**(3), 1121–1127 (2013)
6. J.H. Jeoung: Solidification/stabilization of dredged sludge with low alkalinity additives and geo-environmental assessment. Kyoto University, Kyoto (2003)
7. Di, S., Hong, T., Ling-Chen, M., Hai-Feng, W., Ji-Feng, D.: Experiment on contaminated sediment solidificated by cement, zeolite and TMT. J. Water Resour. Water Eng. (3), 60–64 +71 (2015)
8. Wei, Z., Cheng, L., Lei, L., Ohki, T.: Solidification/stabilization (S/S) of sludge using calcium-bentonite as additive. Environ. Sci. (5), 1020–1025 (2007)
9. Jan-Min, H.: Solidification/stabilization of heavy metal containing sediments using concrete additive. Hunan University (2012)
10. Ansari Mahabadi, A.: Soil cadmium stabilization using an Iranian natural zeolite. Geoderma **137**, 388–393 (2007)

11. Moirou, A., Xenidi, A., Paspaliaris, A.: Stabilization of Pb, Zn, and Cd-contamianted soils by means of natural zeolite. Soil Sediment Contam. **10**, 251–267 (2001)
12. Ashmawy, A.M., Ibrahim, H.S., Abdel Moniem, S.M., Saleh, T.S.: Immobilization of some metals in contaminated sludge by zeolite prepared from local materials. Toxicol. Environ. Chem. **94**(94), 1657–1669 (2012)
13. Xue-Cai, Z., Chang-Shun, H., Shou-Long, L., Rui, F.: Effects of cement solidification on heavy metal extraction from polluted pond sludge in qingshuitang region. J. Cent. S. Univ. Forest. Technol. (4), 150–152 (2009)
14. Hua-Lin, X., Li-Bo, L.: Experimental study on adsorption capability about the heavy metal Ions from water using modified clinoptilolite. Nonmet. Min. (1), 47–49 (2005)
15. Li-Sheng, Z., Shilong, W., Hong, Z., Wei-Xu, G.: Experimental study on the treatment of chromium-containing waste water with zeolite. Environ. Eng. (3), 13–15 (1997)
16. Hui-Sheng, S., Yan-Hong, L.: Research on adsorption of Cr(VI) Ion by modified zeolite. J. Build. Mater. (4), 408–411 (2006)
17. Jun-Zhi, H.: The study on the specification of waste in cement and the Infusion of harmful metal ion. Zhengzhou University (2007)
18. Yu-Shun, G., Yuan-He, L., Nai-Qian, F., Hui-Zhen, L.: The expansion mechanism of clinoptilolite light aggregate. J. Chin. Ceram. Soc. (1), 12–18+129 (1985)
19. Shangbo, Z.: Study on the test system of in-pipe mixing process to solidify the dredged marine sediments and numerical simulation by CFD. Zhejiang Ocean University (2015)
20. Xiao-Peng, H., Zhong-Hui, Z.: Experimental study of drying shrinkage of solidified dredged material. J. Hohai Univ. (Nat. Sci.) **43**(1), 39–43 (2015)
21. De-Hai, Y., Jian-Bing, P.: Experimental study of mechanical properties of chlorite schist with water under triaxial compression. Chin. J. Rock Mechan. Eng. **28**(1), 205–211 (2009)

Performance Evaluation of Low Volume Rural Roads- A State-of-the-Art Review

Piyush G. Chandak[1], Ravindra P. Patil[2], Anand Tapase[3(✉)], Abdulrashid C. Attar[4], and Sabir S. Sayyed[1]

[1] Visvesvaraya Technological University, Belgaum, India
{chandak.p.88,sssabirsayyad}@gmail.com
[2] Jain AGMIT, Jamkhandi, India
ravindrappatil@gmail.com
[3] Karmaveer Bhaurao Patil College of Engineering, Satara, India
tapaseanand@gmail.com
[4] Rajarambapu Institute of Technology, Sakhrale, India
abdulrashid.attar@ritindia.edu

Abstract. The paper provides a state of the art review of performance evaluation of low volume rural roads. Low volume roads are designed based on empirical approach and past experience. The main factors which influence the behavior of a road are tyre pressure and loading intensities, surface and sub-surface temperature, environmental conditions, thickness combinations of different layers, material properties, seepage etc. Consequently, the existing pavement design procedures and analytical procedures are discussed and questioned. Low volume rural roads are not meant for overloaded heavy vehicles, hence are getting deteriorated much earlier than its design life. Secondly, due to the absence of timely maintenance and budgetary provisions for maintenance, problems are increasing resulting in increased number of road fatalities. A combined result of different real conditions in the field affecting the performance of low volume rural roads needs to be studied comprehensively. This paper attempts to study the various methodologies adopted for predicting the behavior of rural roads for heavy traffic and environmental conditions. Also, the importance of the application of finite element method for predicting the behavior of rural roads is showcased. Therefore, it is discussed in details about the need for using an analytical tool like finite element method which will improve the possibility to strengthen the road by the application of a variety of material and thickness combinations for low volume roads.

Keywords: Low volume roads · Finite element method

1 Introduction

In India, low volume roads (LVRs) cover 75% of the overall road system of the nation and are defined as a road, which is negotiable during all weathers, except at major river crossings (MoRTH 2010). It is a commonly observed problem wherein to avoid toll booths and its fees, many vehicles prefer the parallel running low volume roads. Due to such overloaded conditions, the design life of low volume road get affected and hence

© Springer International Publishing AG, part of Springer Nature 2019
W. Frikha et al. (eds.), *New Developments in Soil Characterization and Soil Stability*,
Sustainable Civil Infrastructures, https://doi.org/10.1007/978-3-319-95756-2_5

deteriorate much earlier requiring heavy maintenance and rehabilitation. The current design procedure in India considers the overloaded condition up to 20%, but it is still observed that these low volume roads get affected considerably. It is necessary to modify the design process as per the present day conditions. It has now become imperative that the design process considers the combined effect of such nonrepetitive heavy loads, environmental conditions, and temperature variations to design a suitable and optimum pavement section.

LVRs are generally built as granular pavements with black topped bituminous surfacing. The constantly increasing challenges faced nowadays in designing the low volume roads is the escalating volume and size of traffic along with the number of overloaded trucks using LVRs, change in climatic conditions, costs and moving towards sustainability (Gupta et al. 2014). It is a major challenge to build and maintain these pavements in a financially viable manner using economical materials and techniques (Ranadive and Tapase 2013).

FEM is a multifunctional tool which analyzes and evaluates the joint effect of above factors and predict the behavioral aspects of low volume roads. The ever-increasing flow of commercial vehicles, heavily overloaded trucks along with changes in environmental factors have been reducing the life of the roads (Ranadive and Tapase 2016; Tapase and Ranadive 2017). In the present paper, different changes required in the design procedures and its application to the current scenario is discussed in details.

2 Literature Review

Coghlan (2000) illustrated some prominent features related to low volume roads and main challenges and opportunities that arise from it. The author states that low volume roads comprise 80% of the transportation network and even though traffic volumes are small, the vehicle loads can be high. The conditions of Low Volume Roads occasionally confine and obscure high traffic volume demand and frequently mix unconventional and highway traffic together. El-shaib et al. (2017) compared AASHTO (1993) and MEPDG design guidelines method and observed that the performance of tested pavement sections was different for both the methods and it goes on increasing with enhancement of traffic level as well as for different climate conditions. (Suleiman and Varma 2007) studied for axle loading combinations considered as axles with single tyres, dual tyres, tandem and tridem axles (Kim and Tutumluer 2008). Seasonal condition variations were modeled through strength parameters of the material. From the observed responses, it was deduced that the thaw seasonal condition affected the pavement enormously in terms of damage articulated in the total permanent deformation when compared with the normal season. In comparison, the hot summer season did not have much consequence on LVR pavement damage because the low-strength asphalt layer is too thin to add much to TPD deformation (Amorim et al. 2014). Gupta et al. (2014) proposed a process to account for permanent deformation growth in granular layers of low volume roads. Certain cases of low volume pavement sections were considered for in situ field tests to measure rutting of granular layers. For computation of failure strains and stresses, finite element method was used for mechanistic analysis using properties of the material of the pavement. Using the calculated vertical strain at top of the sub-grade, design charts were developed for traffic up to 1 msa as well as for the value of subgrade modulus between 20 and 150 MPa. The author states that the study could be further extended to include the effect of the various types of sub-grades and aggregates along with the effect of seasonal and weather conditions. Ahmed and Reddy (2012) evaluated the pavement performance for low volume rural roads in terms of rut depth and roughness. The analysis was conducted using elastic layered theory (Kumar et al. 2008) and critical strains for subgrade were also calculated. On the basis of field survey, critical pavement conditions were recognized. The critical condition of the rural roads of the thin surface of the bituminous layer for rut depth was considered as 25 mm, longitudinal depression as 7.3 mm and for roughness 8.5 m/km. Traffic was predicted analogous to the recognized critical conditions of rutting and roughness. Traffic required to cause recognized levels of distress were connected to the vertical subgrade strain values. Design charts were developed using these criteria for low volume roads.

Kalliainen et al. (2015) studied the implications of different tyre configurations of heavy vehicles on a low volume road to measure its performance. PLAXIS 3D 2012 software was deployed to compare the rutting and stress-strain relationships between test site results and modeled version in the software. The comparative analysis showcased that the measured surface deflections and vertical stresses values for the test site and modeled results were within range. High shear strain values were observed in the model for the base course material and the top layers of the subgrade corresponding to the rutting observed on the test site. By performing sensitivity analysis, the authors deduced that the stress and strain levels could be reduced in the base course and subgrade by increasing the thickness of pavement.

Canon Falla et al. (2017) used various test methods to portray the behavior of unbound granular materials used in base and sub-base layers subjected to heavy traffic loads. Based on the conducted test results and use of high-quality UGM, the author suggests that there is a possibility to build pavements with the thickness of asphalt layers less than 50 mm. Abbas et al. (2014) suggested that flexible pavements should be designed as to the ranges of site-specific axle load. Wang and Roque (2011) investigated the pavement response in the close vicinity of the surface for different truck tyre types.

Wu et al. (2011) questioned the assumption that the cementitious stabilized layers do not add to the total permanent deformation of the pavement. The proposed permanent deformation model simulated the behavior of cementitious stabilized materials in a flexible pavement and was incorporated in finite element ABAQUS program. The results obtained from the permanent deformation results from the 6 test sections were used to calibrate the proposed finite element model. Differences in conditions between the laboratory permanent deformation test and pavement under traffic load were accounted for by obtaining a shift factor of 1.13. This data was then utilized to predict rutting depths of selected low volume road sections. The authors state that further research is necessary to develop a model that can forecast the permanent deformation growth in a cementitious stabilized material layer; which affects considerably to the surface rutting.

Ranadive and Katkar (2010), Ranadive and Tapase 2016) used the axisymmetric analysis of flexible pavement using ANSYS and various parameters of pavement were considered for varying thickness conditions. It was deduced that the utmost deflection of flexible pavement varied with an increase in thickness of the various layers of the component while increased when the base course and sub-base course thickness increased. Analytical methods can foresee the performance of flexible pavement and hence expensive field experiments can be avoided (Ranadive and Tapase 2016).

Siddharthan et al. (2005) checked for pavement responses for different loadings and environmental conditions by various bulky off-road vehicles and made the comparative analysis with actual behavior and computed the response of pavement. The authors developed a finite layer method which is used to deal with intricate surface loading along with multiple loads. Kolisoja et al. (2013) developed an approach to calculate the maximum load haulage ability of soft subgrade soils by back calculation of the results obtained from 3-D finite element modeling.

Zuo et al. (2007) studied the outcome of varying water content and temperature on the life of a pavement. The study showed that seasonal temperature variation, temperature gradient and moisture percentage in base and subgrade affected the assessment of pavement life (Diefenderfer et al. 2006). Chiasson et al. (2008) used a two-dimensional finite difference method to compute the temperature allocation in asphalt pavements. Huang (2008) illustrated that both seepage and temperature influence elastic moduli of the different layers of the pavement.

Tighe et al. (2008) analyzed the performance of low volume roads at 6 sites in Canada based on mechanistic-empirical pavement design guide to measure the effects of climatic changes. From the study, it was seen that rutting, and both longitudinal, as well as alligator cracking, would worsen by the change in climate, while transverse cracking would become the lesser amount of trouble. For such predicament; maintenance along with rehabilitation and reconstruction would become essential prior to the design life. For regional baseline climate differences and amplified future traffic, climatic change effects were found to be self-effacing. Potential climate change should be considered in the creation and assessment of prospected designs and maintenance programs (Sahaa et al. 2014). The authors suggest that climatic history and road weather observations for more than 30 years should be thought of in the analysis of pavement deterioration (Mohd Hasan et al. 2015).

Suleiman and Varma (2007) derived a logical approach to establish and make a comparative study of pavement responses for various axle loadings of the truck in relation to normal, thaw and summer season conditions on paved low volume roads. To analyze axle and truck damage factors, responses of pavement in the form of complete permanent deformation was modeled using a finite element program. Howard and James (2014) developed design guidelines for bottom-up fatigue cracking using finite element modeling to consider temperature effects along with the strain measurements obtained from field data.

Henning et al. (2014) studied the influence of climatic conditions (Maadani and Abd El Halim 2017) under various traffic loading conditions and relative factors causative towards the deterioration of low-volume roads. The study revealed that the condition and existence of drainage where required proved to be more imperative rather than only the environmental conditions. Observations made from the results displayed that the rate of rutting for low volume roads was 2.5 times higher on sections with poor drainage in comparison with sections where sufficient drainage was provided. It was also acknowledged that for sections with insufficient drainage, deterioration will be much quicker under heavy volumes of traffic. Bazi et al. (2015) studied the effects of thawing on low volume roads and developed seasonal modification factors for various pavement sections. The study assessed the progression of thawing upon its starting and also the return to equilibrium state during summer. The lower layers of sub-grade continued to thaw while the sub-grade and unbound layers started to recover due to the drying process. The study showcased that during the process of thawing, the sub-grade and base course layers attained 30–40% and 50–70%, respectively, their reference summer time moduli. These ratios were found to be lower than the current design practice. The author recommends a more extensive study to determine the effects of freezing on pavements.

Sinha et al. (2014) demonstrated the helpfulness of finite element method for evaluating the flexible pavement performance (Tapase and Ranadive 2017) using various local materials in the sub-base layer. Industrial waste materials like Linz-Donawitz slag, granulated blast furnace slag (Ziari and Khabiri 2010) and fly ash were used. The research used multilinear elastoplastic hardening model in ANSYS to evaluate the pavement performance considering different types of sub-base materials. Ebrahimi et al. (2012) recommend the use of recycled materials to decrease the thickness of base course by 30% and even augment the life of the pavement by 20%. Kranthi Kumar et al. (2014) studied the various properties of reclaimed asphalt pavement (RAP) for its use in various layers of flexible pavement. The analysis was done for the proportion in between 0 and 40% where it was found that the 20% proportion was optimum when it was compared for different properties like resilient modulus of bituminous concrete mixes, phase angle, bottom top cracking etc. The analysis showed that RAP can be effectively used in surface and base layers in cold and hot bituminous mixes (Vargas-Nordcbeck and Timm 2012).

Saride et al. (2015) recommended the use of higher proportion of reclaimed asphalt pavement (RAP) as a material for the base layer in low volume roads. It was observed that the 80:20 design mix of RAP:Virgin Aggregates with fly ash volume of 40% was found to be an optimum mix and complied with the specified design requirements of the Ministry of Rural Development (MoRD). For this new mix, rutting and fatigue strains were found to be within the permissible limits and the authors reported a reduction of 50% in thickness of the base layer compared with the conservative design mix. Kumar and Akhtar (2016) suggested various locally available materials for sub-base layer in the flexible pavement. Various types of locally available materials were recognized for potential use in sub-base layers with an aim to enhance sub-grade strength. From the Koderma district of Jharkhand, India; construction materials such as gravel, kankar, moorum, construction, and demolition waste were obtained. Various tests were conducted for measuring CBR values, strength properties to attain optimal conditions. As per IRC code, the design of pavement using non-conventional materials improved the subgrade strength and reduced overall pavement thickness by 50 mm. Rodezno and Kaloush (2011) utilized the asphalt-rubber mixes in the design guidelines laid by the MEPDG. Lee et al. (2010) focused on the use of wastes and green approach method to reduce carbon emissions (Yhong et al. 2015).

de Rezende et al. (2015) conducted laboratory and field study to identify the use of non-conventional materials as a substitute for naturally available granular material to be utilized in sub-base and base layers of a pavement. Of the various material combinations, soil–lime mixture along with the lateritic gravel gave better results for both field and laboratory tests. The same behavioral pattern was observed for soil-crushed stone and soil–quarry waste mixtures. It was also seen that quarry waste could replace lateritic gravel in the mixture with lateritic clay for low volume traffic conditions. Such material combinations can hence provide a database for constructing economical, durable and sustainable roads.

Dawson et al. (2007) identified the causes of rutting and based on repeated triaxial load test suggested an analytical approach for road engineers. The data obtained from the test results were incorporated in finite element program FENLAP and analyzed as nonlinear axis-symmetric (Hornych and Salasca 2002). The proposed design strategy does not consider the effect of seasonal frost and does not account for the durability of aggregates. The method is based on granular material testing and for usage by local engineers having limited access to advanced technical resources. Park and Lyttol (2004) showcased the accuracy of using nonlinear analysis for low volume roads using finite element method and applied a single tyre load of 40 KN (Das and Pandey 1999) for the analysis of side boundary. Sobhan (2017) investigated for the rehabilitation of pavement constructed over problematic soils and suggests a correlation of site properties with the performance of the pavement.

Sahoo and Sudhakar Reddy (2010, 2011) attempted to develop a mechanistic-empirical analysis criterion for low volume roads based on the performance results of selective road sections in India. The study is derived from the assumption that the granular material displays non-linearity (Park and Lytton 2002) and stress hardening character. From the results obtained for the selected sections, a rutting depth of 25 mm and roughness index of 8.5 m per km was stated as the terminal conditions for low volume roads. However, the authors state that more extensive work is necessitated to be carried out in other parts to make the research more coherent. Masad et al. (2006) compared the analysis of utilizing anisotropic and isotropic models in flexible pavement response in conjunction with an assessment of fatigue cracking and permanent deformation utilizing (NCHRP 1-37A 2007) design guide.

Qiu et al. (2000) analyzed the structural responses of flexible pavements using finite element method program ARKPAVE. The authors stated that the current design method of using resilient modulus as the lone input parameter is not appropriate and suggests a new design methodology. Based on repeated load testing, a model to predict permanent deformation of sub-grade soils was created and planned a new design criterion for subgrade soils. Duncan et al. were the first researchers to apply the finite element method for analysis of the flexible pavements (Das 2015; Chakroborty and Das 2003). Helwany et al. (1998) showed the helpfulness of finite element method and the use of axis-symmetric 2-D and 3-D analysis. Yoder and Witczak (1975) recommended the use of Boussinesq theory to calculate the strains, deflections and subgrade stresses when the stiffness of the base and subgrade are similar. Saad et al. (2005, 2006) utilized three dimensional finite element model analysis using ADINA program and elasto-plastic models.

In the analysis using finite element method, discretization of continuum into finite elements is performed, above which the governing relationships and properties of materials are applied and articulated in terms of unidentified values at corners. By taking into account the above continuum with suitable loading and constraints, set of equations are formed. The estimated performance of the continuum considered for analysis is found from the solutions of these equations (Zienkiewicz and Taylor 1991; Desai and Abel 2005).

Wanyan et al. (2015) analyzed low volume roads using finite element analysis to calculate the progression of longitudinal shrinkage cracking using moisture content and soil index properties. Various soil samples were collected from 6 clayey sites in Texas, USA for laboratory tests. The authors state that the moisture content based approach provides more flexibility to comprise diverse wetting or drying paths into numerical modeling by laboratory related material constitutive models than the current suction related approach. It was concluded that by enhancing mechanical properties of sub-grade and reducing the moisture fluctuations, on the whole performance of low volume roads constructed over expansive soils can be improved.

Zhao and Dennis (2007) developed a design procedure specifically for low volume roads related to plastic deformations using soil index properties. More than 500 specimens of 20 different subgrade soil samples were considered for obtaining these correlations. Using KENLAYER software, the models for permanent strain prediction and vertical stress prediction were developed dependent on test results of repeated load triaxial testing. The authors have further stated that the developed design procedure has shortcomings mainly related to the inclusion of fatigue cracking, validation of the model with field results, and consideration of variation with time for parameters like temperature, moisture content, and loading conditions.

Bhatore and Tare (2014) developed a system to prioritize the roads network for maintenance. On the basis of dependent variables, performance criteria concerning to the pavement distresses like roughness, rut depth, drainage, and cracking were developed. These equations predicted the pavement performance over a period of time and for a range of traffic and environmental conditions. Data for rural roads was obtained from the districts of Dhar, Jhabua, and Indore of the state of Madhya Pradesh, India. The study recognized rutting, edge drop, cracking and roughness as main dis-tresses. The performance equations were validated with the composed data and can be used in Pavement Maintenance Management System. Nahla et al. (2017) developed a model to predict rutting progression for low volume roads considering various factors like climate, strength of pavement, traffic loading and drainage conditions.

Gupta et al. (2011) proposed a pavement deterioration model which relates to the extrinsic time factor and a combination of intrinsic factors. Sections of low volume pavements from Uttarakhand and Uttar Pradesh, India were measured for the functional and structural response. Artificial neural network and statistical analysis tools were used to develop the models. For improved statistical accuracy, parameters like age of the pavement, traffic, CBR of subgrade and thickness of pavement were considered. Deflection, riding quality, and traffic were the parameters used to develop the main-tenance priority index. The developed models were validated for their use in field conditions. Gupta et al. (2010) investigated the strength properties of the subgrade layer of pavements. Extensive field tests like field dry density, moisture content, dynamic cone penetration and California bearing ratio were performed on low volume road sections from the states of Uttarakhand and Uttar Pradesh, India. Analytical software UK DCP 3.1 was used for calculating pavement strength and designing low volume sealed roads. CBR based method for designing LVRs is used by Indian Roads Congress (IRC) in India. In this study, a connection between DCP Index, CBR and Modified Structural Number (MSN) values was established which could be used to estimate the CBR and MSN value from the DCP data for low volume roads.

Tawalare and Raju (2016) presented a methodology to measure pavement performance index for rural roads. Distress parameters for rural roads and rating criteria were recognized through literature review and opinions of experienced industrial experts. The distress parameters identified for Indian rural roads were skid resistance, unevenness, CBR, potholes, drainage conditions, raveling, rut depth, camber, carriage width and thickness, patching, edge break, cracking, the condition of shoulder and age. For each parameter causing distress, weight for extremity was calculated based on data of questionnaire survey of professionals working in Pradhan Mantri Gram Sadak Yojana, India. The Pavement performance index can be used to prepare precedence list of rural roads for repair and maintenance schedule.

Avinash et al. (2014) evaluated low volume urban roads in Karnataka, India for volume of traffic, functional as well as structural condition in relation to CBR value of the sub-grade, deflection survey, International Roughness Index (IRI) and conditional rating dependent on the amount of cracking, rutting and potholes. To calculate critical strains KENLAYER software was used. For the chosen road network, the methods of pavement condition index and maintenance priority index did not yield reasonable results and hence, modified maintenance priority index was proposed to give a practical ranking of the necessity of maintenance priority. The authors state that the research could further be extended to make it more comprehensive.

3 Current Design Procedures

The earlier design approach was based on various design procedures of relevant IS codes of Indian Roads Congress, Ministry of Road Transport and Highways Specification for road and bridge works and other documents. IRC: SP: 20-2002, rural roads manual was the initial guidelines procedure adopted in India specifically for rural roads including Other than district roads and village roads having a low volume of traffic. Pavement design was evaluated on the basis of commercial vehicles per day (CVPD) combining HCV, MCV, and LCV together without considering overloaded vehicles traveling on rural roads (MoRTH 2010).

Based on CVPD, four traffic categories were constituted as 0–15, 15–45, 45–150 and 150–450 CVPD; pavement design curves were provided with minimum base course thickness of 150 mm for curves A and B while 225 mm for curves C and D. Sub-base course thickness was calculated by subtracting minimum base course thickness from the required total thickness of the pavement. There is no separate provision for design procedure of unsealed gravel roads in this manual (IRC: SP: 20-2002).

IRC: SP: 72-2007, (Guidelines for the Design of Flexible Pavements for Low Volume Roads) is the revision of Chap. 5 of IRC: SP: 20-2002 which relates to guidelines for the design of flexible pavements for low volume rural roads. For low volume roads, the equivalent standard axle load (ESAL) is considered up to 1 million standard axles (msa). For flexible pavement design, traffic over 1,000,000 ESAL, IRC:37 is to be considered. In this design manual, the low volume rural roads were divided into three categories namely: gravel roads, flexible pavements, and rigid pavements. The design procedure adopted in this manual was performance-based derived from experiences in USA on low volume road design, as depicted in AASHTO guide for the design of pavement structures (AASHTO 1993; IRC: 37-2012).

As per the guidelines, design traffic parameter in terms of cumulative standard axle load of 80 KN is considered for the design life of 10 years with due consideration for enhanced traffic during harvesting season. For subgrade strength evaluation, moisture content has been considered rather than always being dependent on 4 days soaked CBR values. The subgrade strength has been classified into 5 groups while traffic has been classified into 7 ranges. The typical cross section of low volume roads consists of sub-base, base, and bituminous surface. The thickness of flexible pavements was based on structural number recommended by AASHTO for low volume roads for USA climatic zones. Generally, the thicknesses of the mentioned layer should not be less than 100, 150 mm for sub-base and base layer respectively.

IRC: SP: 72-2015 (Guidelines for the Design of Flexible Pavements for Low Volume Roads) is the first revision of IRC: SP: 72-2007 which considers design charts for traffic more than 1 msa and up to 2 msa. It has made significant changes in the design approach wherein the equivalent standard axle load (ESAL) is now considered up to 2 million standard axles (msa) and traffic categories T_8 and T_9 have been included in the design catalog. The design charts in this manual now consider the strength of sub-grade and design traffic which was not available in the previous design guidelines.

As per the new guidelines mentioned in the latest design code of IRC: SP: 72-2015, the design traffic parameter in terms of cumulative standard axle load of 80 KN is considered for the design life of 10 years. For subgrade strength evaluation, moisture content has been considered rather than always being dependent on 4 days soaked CBR values. It is also suggested that all rural roads should be designed to a sub-grade of minimum 5%. The sub-grade strength has been classified into 5 groups while traffic has been classified into 9 ranges, considering the ESAL up to 2 msa. The typical cross section of low volume roads as suggested by the code is of soil-cement stabilized sub-base, base and the top layer of premix carpet or two layer surface dressing.

For traffic up to 60,000 ESAL and sub-grade CBR between 2 and 5, gravel roads perform satisfactorily. While sub-grade CBR is above 5, gravel roads are recommended for traffic up to 1,00,000 ESAL applications. For traffic above 1,00,000 ESAL applications the guidelines recommend a black topped flexible pavement with lime/cement soil stabilized layers for silty or clayey soils. The minimum thicknesses specified for granular or lime/cement soil stabilized layers are 100 mm for sub-base, 150 mm for base layer, and bituminous layer (BL) comprising of 50 mm bituminous macadam and 20 mm premix carpet as surface dressing. The thickness of 2500 mm can be adequately assumed for sub-grade Gupta et al. (2014). The material used for shoulder should comply with the quality of sub-base material used and compacted to a minimum thickness of 100 mm.

4 Observations and Future Research Direction

Considering the available literature and the design guidelines for low volume rural roads it is observed that considerable progress has been made in the design procedure bearing in mind many factors which were previously ignored. It is seen that substantial progress has been made in relation to the use of finite element models, multilayered elastic theories etc. Due consideration has also been given to low volume rural roads which can be designed for flexible as well as rigid pavements (Douglas 2016) and includes design for traffic load up to 2 msa inclusive of overloaded vehicles. However, there are some factors which still need to be addressed in the designing process of low volume rural roads.

The design charts based on different values of CBR should be tested for thickness variation of different layers of the pavement section of low volume roads. An optimum design mix can be obtained by evaluating the pavement section for variation in thicknesses. As such keeping in mind both the important parameters like initial costs, maintenance costs, design life a suitable design can be suggested for different values of CBR. The design procedure is extensively drawn based on the findings of AASHTO for the USA climatic zones with consideration of reliability level but it is imperative to consider the different climatic conditions that affect the pavement for various regions. The effect of the groundwater table, frost and thaw action, seepage pressure need to be given appropriate thought rather than just increasing the thickness of the pavement section. Further, the study on climatic conditions can well be extended to consider the effect of daytime temperature and nighttime temperature on the pavement and also due to seasonal changes along with decremental and incremental variation within the different component layers. The life of the pavement decreases due to seepage through sub-grade while the proximity of water table on swelling as well as shrinkage property of sub-grade also needs to be studied in details.

From the literature, it is observed that fatigue and rutting are the main factors considered for determining the deterioration of pavement and is theoretically linked to CBR values of the sub-grade. It is necessary to calculate the life of pavement in the form of a number of cumulative standard axles and then it can be associated with the rutting life and fatigue life. The strength characteristics are determined depending on the various material constituents for each layer. Hence, it is necessary to investigate the effect of varying thicknesses of each layer on the performance of a low volume road. Further, the 2-D finite element models in conjunction with plain strain, axis-symmetric, and plain stress are utilized successfully for the assessment of critical pavement responses. A suitable blend of 2D and 3D finite element analysis can help in obtaining the realistic results and avoid heavy computational time that will be required (Chandak et al. 2017). To illustrate the performance of the materials in different layers of the pavement as per different environmental conditions and requirements, material constitutive models like linear-elastic, elastoviscoplastic, visco-elastic and nonlinear-elastic (Gonzalez et al. 2007; Huang 2008) models should be used.

From the study of various guidelines, it is observed that the use of waste materials and locally available materials is recommended but there is a lack of detailed guidelines as to use of these materials to make the pavement section more sustainable.

Incorporation of naturally available waste materials as well as locally available materials can be studied in details and can be analyzed for its optimum use in the pavement section. From the literature study, it is also seen that reclaimed asphalt pavement has been successfully utilized in the pavement section and hence this study should further be extended to use of stress absorbing membrane interlayer in low volume roads.

Although the design guidelines have considered the effect of overloaded vehicles up to 20% for HCV and MCV it is observed in reality that this value exceeds much more than that and hence can be checked for such overloaded conditions. Due to such overloaded conditions, it is also necessary to investigate the high tyre pressures exerted on the pavement of low volume roads. Considering all the above factors and literature it is mostly seen that only one parameter at a time is considered and hence there is a need to study the collective effect of these parameters together on the performance of a pavement. A more rigorous study and development of such analytical tool will help give a more realistic approach to the design of low volume roads.

References

Abbas, A.R., Frankhouser, A., Papagiannakis, A.T.: Effect of traffic load input level on mechanistic—empirical pavement design. Transportation Research Record: Journal of the Transportation Research Board, No. 2443, Transportation Research Board of the National Academies, Washington, DC, pp. 63–77 (2014). https://doi.org/10.3141/2443-08

Ahmed, S., Reddy, K.S.: Performance criteria for rural road pavements. In: Proceedings of International Conference on Advances in Architecture and Civil Engineering (AARCV 2012) (2012)

American Association of State Highway Officials AASHTO: AASHTO guide for the design of pavement structures. AASHTO, Washington, DC (1993)

Amorim, S.I.R., Paisa, J.C., Valea, A.C., Minhotob, M.J.C.: A model for equivalent axle load factors. Int. J. Pavement Eng. (2014). https://doi.org/10.1080/10298436.2014.968570

Avinash, N.R., Vinay, H.N., Prasad, D., Dinesh, S.V., Dattatreya, J.K.: Performance evaluation of low volume flexible pavements—a case Study. T&DI, ASCE, 69–78 (2014)

Bazi, G., Briggs, R., Saboundjian, S., Ullidtz, P.: Seasonal effects on a low-volume road flexible pavement. Transportation Research Record: Journal of the Transportation Research Board, No. 2510, Transportation Research Board, Washington, DC, pp. 81–89 (2015). https://doi.org/10.3141/2510-10

Bhatore, A., Tare, V.: Performance models for rural roads. Indian Highw. 42(2), 82–88 (2014)

Canon Falla, G., Leischner, S., Blasl, A., Erlingsson, S.: Characterization of unbound granular materials within a mechanistic design framework for low volume roads. Transp. Geotech. (2017). https://doi.org/10.1016/j.trgeo.2017.08.010

Chakroborty, P., Das, A.: Principles of Transportation Engineering. Prentice Hall of India Private Ltd., New Delhi (2003)

Chandak, P.G., Tapase, A.B., Sayyed, S.S., Attar, A.C.: A State-of-the-Art Review of Different Conditions Influencing the Behavioral Aspects of Flexible Pavement. Advancement in the Design and Performance of Sustainable Asphalt Pavements, pp. 300–312. Springer International Publishing (2017). https://doi.org/10.1007/978-3-319-61908-8_22

Chiasson A., Yavuzturk C., Ksaibati K.: Linearized approach for predicting thermal stresses in asphalt pavements due to environmental conditions. J. Mater. Civil Eng. ASCE 20(2), 118–127 (2008). https://doi.org/10.1061/(asce)0899-1561(2008)20:2

Coghlan, G.T.: Opportunities for low-volume roads. Transportation Research Board CD (2000)

Das, A.: Analysis of pavement structures. CRC Press, Taylor and Francis, Boca Raton (2015)

Das, A., Pandey, B.B.: The M-E design of bituminous road: an indian perspective. J. Transp. Eng. ASCE **125**(5), 463–471 (1999). https://doi.org/10.1061/(asce)0733-947X(1999)125:5(463)

Dawson, A.R., Kolisoja, P., Vuorimies, N., Saarenketo, T.: Design of low-volume pavements against rutting. Transp. Res. Rec. J. Transp. Res. Board, 165–172 (2007). https://doi.org/10.3141/1989-19

de Rezende, L.R., Marques, M.O., da Cunha, N.L.: The use of nonconventional materials in asphalt pavements base. Road Mater. Pavement Des. **16**(4), 799–814 (2015). https://doi.org/10.1080/14680629.2015.1055334

Desai, C.S., and Abel, J.F.: Introduction to the Finite Element Method—A Numerical Method for Engineering Analysis. CBS Publishers and Distributors, Delhi (2005)

Diefenderfer, B.K., Al-Qadi, I.L., Diefenderfer, S.D.: Model to predict pavement temperature profile: development and validation. J. Transp. Eng. ASCE **132**(2), 162–167 (2006)

Douglas, R.A.: Low-Volume Road Engineering: Design, Construction, and Maintenance. CRC Press, Taylor & Francis Group, Boca Raton (2016)

Ebrahimi, A., Kootstra, B.R., Edil, T.B., Benson, C.H.: Practical approach for designing flexible pavements using recycled roadway materials as base course. Road Mater. Pavement Des. **13**(4), 731–748 (2012). https://doi.org/10.1080/14680629.2012.695234

El-Shaib, M.A., El-Badawy, S.M., Shawaly, E.-S.A.: Comparison of AASHTO 1993 and MEPDG considering the Egyptian climatic conditions. Innov. Infrastruct. Solut. Springer International Publishing, Switzerland **2**, 18 (2017). https://doi.org/10.1007/s41062-017-0067-6

Gonzalez, A., Saleh, M.F., Ali A.: Evaluating nonlinear elastic models for unbound granular materials in the accelerated testing facility. Transportation Research Record No. 1990, pp. 141–149 (2007)

Gupta, A., Kumar, P., Rastogi, R.: Field and laboratory investigations on subgrade layer of low volume roads. In: Indian Geotechnical Conference, pp. 1019–1022 (2010)

Gupta, A., Kumar, P., Rastogi, R.: Mechanistic–empirical approach for design of low volume pavements. Int. J. Pavement Eng. Taylor & Francis (2014). https://doi.org/10.1080/10298436.2014.960999

Gupta, A., Kumar, P., Rastogi, R.: Pavement deterioration and maintenance models for low volume roads. Int. J. Pavement Res. Technol. **4**(4), 195–202 (2011)

Helwany, S., Dyer, J., Leidy, J.: Finite element analysis of flexible pavement. J. Transp. Eng. ASCE, **124**(5), 491–499 (1998). https://doi.org/10.1061/(asce)0733947X(1998)124:5(491))

Henning, T.F.P., Alabaster, D., Arnold, G., Liu, W.: Relationship between traffic loading and environmental factors and low-volume road deterioration. Transp. Res. Rec. J. Transp. Res. Board, 100–107 (2014). https://doi.org/10.3141/2433-11

Hornych, P., Salasca, S.: Prediction of the behavior of a flexible pavement using finite element analysis with non-linear elastic and viscoelastic models. In: Proceedings of International Conference on Asphalt Pavements, Denmark (2002)

Howard, I., James, R.: Bottom-up fatigue cracking of low-volume flexible pavement analysis from instrumented testing. Transp. Res. Rec. J. Transp. Res. Board (2014). https://doi.org/10.3141/2094-05

Huang, Y.: Pavement Analysis and Design, 2nd edn. Pearson Education Inc. and Dorling Kindersley Publishing Inc., New York (2008)

IRC: 37-2012: Guidelines for the design of flexible pavements. Indian Roads Congress, New Delhi

IRC: SP: 20-2002: Rural roads manual. Indian Roads Congress, New Delhi

IRC: SP: 72-2007: Guidelines for the design of flexible pavements for low volume rural roads. Indian Roads Congress, New Delhi

IRC: SP: 72-2015 (First Revision): Guidelines for the design of flexible pavements for low volume rural roads. Indian Roads Congress, New Delhi

Kalliainen, A., Kolisoja, P., Haakana, V.: Effect of tyre configuration on the performance of a low-volume road exposed to heavy axle loads. Transp. Res. Rec. J. Transp. Res. Board, 174–184 (2015). https://doi.org/10.3141/2474-21

Kim, M., Tutumluer, E.: Implications of complex axle loading and multiple wheel load interaction in low volume roads. Airfield Highw. Pavements, ASCE (2008)

Kolisoja, P., Kalliainen, A., Vuorimies, N.: Mechanistic design of low volume road structures. In: Hoff, I., Mork, H., Saba Garba, R. (eds.) Proceedings of the Ninth International Conference on the Bearing Capacity of Roads, Railways and Airfields (BCRRA), Norway, vol. 1, pp. 331–340. Akademika Publishing, Norja (2013). International Conference on the Bearing Capacity of Roads, Railways and Airfields (1)

Kranthi Kumar, K., Rajasekhar, R., Amarnatha Reddy, M., Pandey, B.B.: Reclaimed asphalt pavements in bituminous mixes. Indian Highw. 12–18 (2014)

Kumar, S., Nazish Akhtar, E.: Use of locally available non-conventional materials in the construction of rural road pavements. Indian Highw. **44**(9) (2016). ISSN 0376-7256

Kumar, S.S., Sridhar, R., Reddy, K.S., Bose, S.: Analytical investigation on the influence of loading and temperature on top-down cracking in bituminous layers. J. Indian Road Congr. **69**(1), 71–77 (2008)

Maadani, O., Abd El Halim, A.O.: Environmental considerations in the AASHTO mechanistic-empirical pavement design guide: impacts on performance. J. Cold Reg. Eng. ASCE (2017). ISSN 0887-381X

Masad, S., Little, D., Masad, E.: Analysis of flexible pavement response and performance using isotropic and anisotropic material properties. J. Transp. Eng. ASCE **132**(4), 342–349 (2006)

Ministry of Road Transport and Highways (MoRTH): Basic road statistics of India 2004–2008, Report by Transportation Research Wing, Government of India, New Delhi (2010)

Mohd Hasan, M.R., Hiller, J.E., You, Z.: Effects of mean annual temperature and mean annual precipitation on the performance of flexible pavement using ME design. Int. J. Pavement Eng. (2015). https://doi.org/10.1080/10298436.2015.1019504

Nahla, A., Rayya, H., Bayar, M.: Multilevel modeling of rutting progression for low-volume roads. Road Transp. Res. J. Aust. New Zealand Res. Pract. **26**(2), 22–35 (2017)

NCHRP: Evaluation of mechanistic-empirical design procedure. National Cooperative Highway Research Program, NCHRP Project 1-37A, National Research Council, Washington (2007)

Park, S.-W., Lyttol, R.L.: Effect of stress-dependent modulus and Poisson's ratio on structural responses in thin asphalt pavement. J. Transp. Eng. ASCE **130**(3), 387–394 (2004). https://doi.org/10.1061/(asce)0733-947x(2004)130:3(387)

Park, S.W., Lytton, R.L.: Prediction of flexible pavement response using non-linear stress-dependent material models. In: Proceedings of the 6th International Conference on the Bearing Capacity of Roads and Airfields, Lisbon, Portugal, no. 1, pp. 231–40, June 2002

Qiu, Y., Dennis, N.D., Elliot, R.P.: Design criteria for permanent deformation of subgrade soils in flexible pavements for low volume roads. Soils Found. Jpn. Geotech. Soc. **40**(10), 1–10 (2000)

Ranadive, M.S., Katkar, A.B.: Finite element analysis of flexible pavements. Indian Highw. **38**(6) (2010)

Ranadive, M.S., Tapase, A.: Pavement performance evaluation for different combinations of temperature conditions and bituminous mixes. Innov. Infrastruct. Solut. Springer **1**(40), 15 (2016). https://doi.org/10.1007/s41062-016-0040-9

Ranadive, M.S., Tapase, A.B.: Investigation of behavioral aspects of flexible pavement under various conditions by finite element method. In: Yang, Q., Zhang, J.-M., Zheng, H., Yao, Y. (eds.) Constitutive Modeling of Geomaterials, pp. 765–770. Springer, Berlin (2013). https://doi.org/10.1007/978-3-642-32814-5_100

Ranadive, M.S., Tapase, A.B.: Parameter sensitive analysis of flexible pavement. Int. J. Pavement Res. Technol. (IJPRT), Elsevier, Spec. Iss. Sustain. Pavement Eng. (2016). https://doi.org/10.1016/j.ijprt.2016.12.001

Rodezno, M.C., Kaloush, K.E.: Implementation of Asphalt-Rubber mixes into the mechanistic empirical pavement design guide. Road Mater. Pavement Des. 12(2), 423–439 (2011). https://doi.org/10.1080/14680629.2011.9695252

Saad, B., Mitri, H., Poorooshasb, H.: Three-dimensional dynamic analysis of flexible conventional pavement foundation. J. Transp. Eng. ASCE 131(6), 460–469 (2005). https://doi.org/10.1061/asce0733-947x(2005)131:6(460)

Saad, B., Mitri, H., Poorooshasb, H.: 3D FE analysis of flexible pavement with geosynthetic reinforcement. J. Transp. Eng. ASCE 132(5), 402–415 (2006). https://doi.org/10.1061/asce0733-947x2006132:5(402)

Sahaa, J., Nassirib, S., Bayatc, A., Soleymanid, H.: Evaluation of the effects of Canadian climate conditions on the MEPDG predictions for flexible pavement performance. Int. J. Pavement Eng. 15(5), 392–401 (2014). https://doi.org/10.1080/10298436.2012.752488

Sahoo, U.C., Sudhakar Reddy, K.: Effect of nonlinearity in the granular layer on critical pavement responses of low volume roads. Int. J. Pavement Res. Technol. 3(6), 320–325 (2010)

Sahoo, U.C., Sudhakar Reddy, K.: Performance criterion for thin-surface low-volume roads. Transportation Research Record: Journal of the Transportation Research Board, No. 2203, Transportation Research Board of the National Academies, Washington, DC, pp. 178–185 (2011). https://doi.org/10.3141/2203-22

Saride, S., Avirneni, D., Chandra Prasad Javvadi, S.: Utilization of reclaimed asphalt pavements in Indian low-volume roads. J. Mater. Civil Eng. ASCE (2015). https://doi.org/10.1061/(asce)mt.1943-5533.0001374

Siddharthan, R.V., Sebaaly, P.E., El-Desouky, M., Strand, D., Huft, D.: Heavy off-road vehicle tyre-pavement interaction and response. J. Transp. Eng. ASCE 131(3), 239–247 (2005). https://doi.org/10.1061/(ASCE)0733-947X(2005)131:3(239)

Sinha, A.K., Chandra, S., Kumar, P.: Finite element analysis of flexible pavement with different subbase materials. Indian Highw. New Delhi 42(2), 53–63 (2014)

Sobhan, K.: Challenges Due to Problematic Soils: A Case Study at the Crossroads of Geotechnology and Sustainable Pavement Solutions. Innovative Infrastructure Solution, pp. 1–18. Springer International Publishing (2017). https://doi.org/10.1007/s41062-017-0070-y

Suleiman, N., Varma, A.: Modeling the response of paved low-volume roads under various traffic and seasonal conditions. Transp. Res. Rec. J. Transp. Res. Board, 230–236 (2007). https://doi.org/10.3141/1989-68

Tapase, A., Ranadive, M.: Performance evaluation of flexible pavement using finite element method. In: ASCE GSP 266, GeoChina 2016: Material, Design, Construction, Maintenance and Testing of Pavement, pp. 9–17 (2016). https://doi.org/10.1061/9780784480090.002

Tapase, A.B., Ranadive, M.S.: Predicting performance of flexible pavement using finite element method. In: Mohammad, L. (ed.) Advancement in the Design and Performance of Sustainable Asphalt Pavements. GeoMEast 2017. Sustainable Civil Infrastructures. Springer, Cham (2017). https://doi.org/10.1007/978-3-319-61908-8_11

Tawalare, A., Raju, K.V.: Pavement performance index for Indian rural roads. Perspect. Sci. (2016). https://doi.org/10.1016/j.pisc.2016.04.101

Tighe, S.L., Smith, J., Mills, B., Andrey, J.: Evaluating climate change impact on low-volume roads in Southern Canada. Transp. Res. Rec. J. Transp. Res. Board, 9–16 (2008). https://doi.org/10.3141/2053-02

Vargas-Nordcbeck, A., Timm, D.H.: Rutting characterization of warm mix asphalt and high RAP mixtures. Road Mater. Pavement Des. **13**(Suppl 1), 1–20 (2012). https://doi.org/10.1080/14680629.2012.657042

Wang, G., Roque, R.: Evaluation of truck tyre types on near-surface pavement response based on finite element analysis. Int. J. Pavement Res. Technol. **4**(4), 203–211 (2011)

Wang, Y., Chong, D.: Long-life flexible pavement: myth, reality, and the way forward. ASCE New Front. Road Airp. Eng. 268–283 (2015)

Wanyan, Y., Abdallah, I., Nazarian, S., Puppala, A.J.: Moisture content based longitudinal cracking prediction and evaluation model for low volume roads over expansive soils. J. Mater. Civil Eng. ASCE (2015). https://doi.org/10.1061/(asce)MT.1943-5533.0001217

Wu, Z., Chen, X., Yang, X., Zhang, Z.: Finite element model for rutting prediction of flexible pavement with cementitiously stabilized base–subbase. Transp. Res. Rec. J. Transp. Res. Board 104–110 (2011). https://doi.org/10.3141/2226-11

Lee, M., Tan Poi Cheong, D., Dong Qing, W.: Green approach to rural roads construction—stabilization of in-situ soils and construction wastes. In: The 7th Asia Pacific Conference on Transportation and the Environment, Semarang, Indonesia

Yoder, E.J., Witczak, M.W.: Principles of pavement Design, 2nd edn. Wiley-Interscience Publication, New York (1975)

Zhao, Y., Dennis Jr., N.D.: Development of a simplified mechanistic–empirical design procedure for low-volume flexible roads. Transp. Res. Rec. J. Transp. Res. Board, 130–137 (2007). https://doi.org/10.3141/1989-15

Ziari, H., Khabiri, M.M.: Preventive maintenance of flexible pavement and mechanical properties of steel slag asphalt. J. Environ. Eng. Landsc. Manag. Taylor and Francis **15**(3), 188–192 (2010). https://doi.org/10.1080/16486897.2007.9636928

Zienkiewicz, O.C., Taylor, R.L.: The Finite Element Method, vol. 2. McGraw-Hill, New York (1991)

Zuo, G., Drumm, E.C., Meier, R.W.: Environmental effects on the predicted service life of flexible pavements. J. Transp. Eng. **133**(1), 47–56 (2007)

Evaluation of the Load Bearing Capacity of Piles by Numerical, Analytical and Empirical Approaches

Karen Ninanya[1(⊠)], Jackeline Huertas[2], Hugo Ninanya[3], and Celso Romanel[3]

[1] Engineer at Essential Material Consulting (EMC), Lima, Peru
kninanya@outlook.com
[2] Geotechnical Engineer at Essential Material Consulting (EMC), Lima, Peru
jcastaneda@emconsulting.pe
[3] Pontifical Catholic University of Rio de Janeiro (PUC-Rio),
Rio de Janeiro, Brazil
hninanya@aluno.puc-rio.br, romanel@puc-rio.br

Abstract. The evaluation of axial bearing capacity of piles plays an essential role in foundation design. It can be estimated by theoretical methods, through an analytical or numerical solution, or by semi-empirical methods, whose application is generally based on field test results. Semi-empirical formulations are often preferred by geotechnical engineers since theoretical approaches are generally limited to particular applications. This paper compares the axial bearing capacity of bored piles estimated by different methodologies, based on instrumented load tests carried out in the Campinas and the Brasilia university campi, in Brazil. The semi-empirical approach involved settlement control and extrapolation of load-settlement curve while the theoretical approach was based on both analytical and numerical solutions, the latter computed through finite element analyses. When compared to the bearing capacity measured in the instrumented load tests, both numerical and the settlement control results showed close agreement, while the extrapolation of the load-settlement curve estimated higher than expected results and the theoretical method presented conservative values when introducing the negative skin friction effects.

1 Introduction

Piles are one of the oldest alternatives to support structures, generally used when the most superficial layers of the soil profile are not sufficiently resistant to support loads from superstructure. However, their design is still a challenge for geotechnical engineers, which often make use of design techniques largely based on empirical principles.

This paper studies the excavation process of bored piles, in saturated and unsaturated cohesive soils, with the aim to compare the bearing capacity estimates to the static ultimate load measured in the field. Three case studies in Brazil were analyzed by different approaches, among them analytical and numerical methods, as well as semi-empirical approaches proposed by O'Neill and Reese (1999), Aoki and Velloso (1975), Monteiro (1977), Décourt and Quaresma (1982), Van der Veen (1953), Brinch-Hansen (1963), Chin (1971), Massad (1986), Mazurkiewicz (1972), Butler and Hoy (1977), Décourt (1996), Davisson (1972) and NBR-6122 (1996).

© Springer International Publishing AG, part of Springer Nature 2019
W. Frikha et al. (eds.), *New Developments in Soil Characterization and Soil Stability*,
Sustainable Civil Infrastructures, https://doi.org/10.1007/978-3-319-95756-2_6

2 Bearing Capacity of Piles

2.1 Axial Bearing Capacity: Analytical Method

In this research, it will be considered the O'Neill and Reese method (1999) to estimate the bearing capacity of a bored pile taking into account the effect of the negative skin friction. It is known that the bearing capacity (P_{ult}) of a pile is the sum of the end bearing resistance (P_b) and skin friction resistance (P_s):

$$P_{ult} = A_p \left(\xi_{sc} \xi_{dc} \xi_{ic} N_c c + \xi_{sq} \xi_{dq} \xi_{iq} \left(N_q - 1 \right) \sigma_{vb} \right) + \sum (r_L)_i (A_s)_i \qquad (1)$$

where A_p is the area of pile tip, N_c and N_q are bearing capacity factors, ξ_{sc} and ξ_{sq} are shape factors, ξ_{dc} and ξ_{dq} are depth factors, ξ_{ic} and ξ_{iq} are correction factors, c is the average cohesion of soil just below the pile tip, σ_{vb} is the effective vertical pressure at the base level, r_L is the average unit skin friction of each individual soil layer and A_s is the area of the bored pile interface with each soil layer. The values of N_c, N_q, ξ_{sc}, ξ_{sq}, ξ_{dc}, ξ_{dq}, ξ_{ic} and ξ_{iq} are calculated in function of the resistance parameters and geometrical characteristics of the pile.

2.1.1 Effect of the Negative Skin Friction

This effect reduces the skin friction resistance significantly. Although it is impossible to predict accurately this effect, different methods are used to estimate it. It is considered in this research the alpha method (Tomlinson 1957, 1971) where the skin bearing capacity includes the negative skin friction effect (P_S'), given by Eq. (2), which replaces the second term (P_s) in Eq. (1).

$$P_s' = \sum (c_a)_i (A_s)_i \qquad (2)$$

where c_a is the adhesion factor.

2.2 Axial Bearing Capacity: Semi-empirical Methods

The semi-empirical methods suggested by Aoki and Velloso (1975) and Monteiro (1977), considering results from the Standard Penetration Test (SPT), are expressed by Eq. (3) while the Decourt-Quaresma method (1982) estimates the pile bearing capacity through Eq. (4):

$$P_{ult} = A_P \frac{k N_P}{F1} + U \sum \frac{\alpha k N_M}{F2} \Delta L \qquad (3)$$

$$P_{ult} = A_P k N_1 + 10 U L \left(\frac{N_2}{3} + 1 \right) \qquad (4)$$

where the empirical factors $F1$ and $F2$ depend on the type of pile, the correction factors k and α vary according to soil characteristics and N_P, N_M, N_1 and N_2 are SPT-N values. In Eq. (4) U and L are the perimeter and the length of the pile, respectively.

2.3   Axial Bearing Capacity: Extrapolation of the Load-Settlement Curve

Methodologies based on the extrapolation of the load-settlement curve adjust the curve to a mathematical function in order to predict both bearing capacity and the corresponding settlement. Van der Veen method (1953) assumes an exponential formulation, as given by Eq. (5), the Brinch-Hansen (1963) and Chin (1971) methods consider a hyperbolic behavior, as shown by Eqs. (6) and (7), respectively, and the Massad method (1986) adopts a polynomial function, described by Eq. (8). On the other hand, Mazurkiewicz (1972) and Butler and Hoy (1977) methods estimate the bearing capacity graphically, drawing straight lines in the load–settlement curve, and the Decourt method (1996) considers a linear variation of the pile stiffness K, as indicated by Eq. (9).

$$P = P_{ult}(1 - e^{\alpha_i r + \beta_i}) \tag{5}$$

$$P = \frac{\sqrt{r}}{\alpha_i + \beta_i r} \tag{6}$$

$$P = \frac{r}{\alpha_i + \beta_i r} \tag{7}$$

$$P_{(n+1)} = \alpha_i + \beta_i P_n \tag{8}$$

$$K = \alpha_i + \beta_i P \tag{9}$$

In Eqs. (5), (6), (7), (8) and (9), αi and βi are the adjustment factors for each method. In Eqs. (8) and (9), n represents a series of values (1, 2, 3, 4, ..., n) that indicate the corresponding applied load and K is the pile stiffness defined as the ratio of the applied load (P) and the settlement (r), respectively.

2.4   Axial Bearing Capacity: Settlement Control

Methods based on settlement control define the bearing capacity in association to a settlement value that exceeds the elastic compression of the pile. The Brazilian Technical Standard NBR-6122 (1996) establishes such value as $(D/30)$ mm while the Davisson (1972) considers $(4 + D/120)$ mm, where D is the pile diameter.

2.5   Axial Bearing Capacity: Finite Element Analysis

In order to calculate the axial capacity of pile under vertical load, a numerical model has been developed using the finite element program Plaxis 2D. The problem was represented as an axisymmetric model, considering a finite element mesh whose width is twice the pile's length and a depth between 1.4 and 1.5 times the foundation's length. The Mohr-Coulomb elastoplastic constitutive model was chosen to represent the mechanical behavior of the soil. The pile was made up of reinforced concrete and the behavior is assumed to be linear-elastic. It is worth mentioning that the load–settlement curves extracted from the numerical results were obtained from a control point located on the top of the pile in order to estimate its bearing capacity.

3   Case Studies

The present research investigates two case studies of instrumented pile load tests carried out in the experimental fields situated in the campus of the universities of Campinas (State of Sao Paulo) and Brasilia (Federal District), in Brazil. The static loading test and SPT-N values were taken from Albuquerque (2001), for the Campinas case, and from Mascarenha (2003), for the Brasilia case. Both cases involve analysis of bored piles in residual soils.

3.1   Campinas Case

3.1.1   Residual Soil Deposit and Parameters

In the Campinas experimental field three cylindrical bored piles were tested (12 m length and 0.4 m diameter) with a measured bearing capacity of 660 kN. Basically, the soil stratigraphy is conformed of 6.5 m of silty-sandy clay followed by 10.5 m of clay-sandy silt and the phreatic level is located 19 m below the ground surface, as shown in Fig. 1. This figure also gives the distribution of SPT-N values along depth, which were correlated to obtain the resistance parameters (c,) and the unit weights, necessary for the bearing capacity calculation, as indicated in Fig. 2.

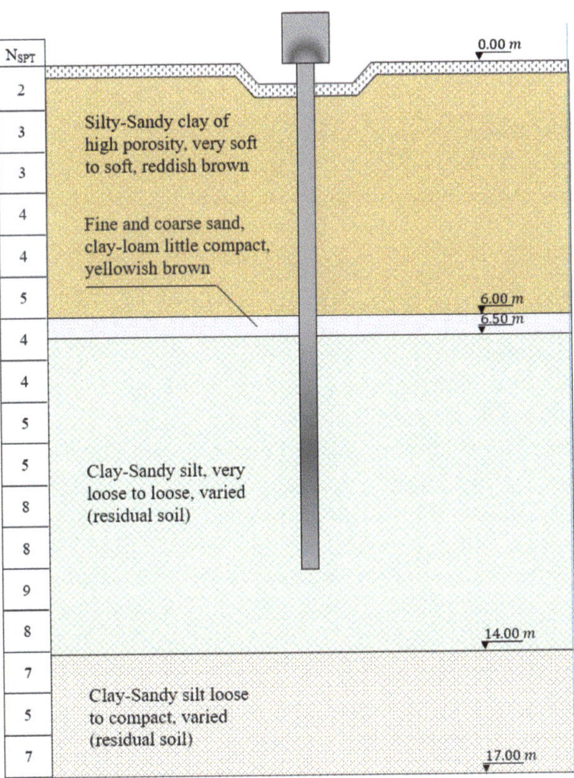

Fig. 1. Pile load test in the experimental field of Campinas (modified from Albuquerque 2001).

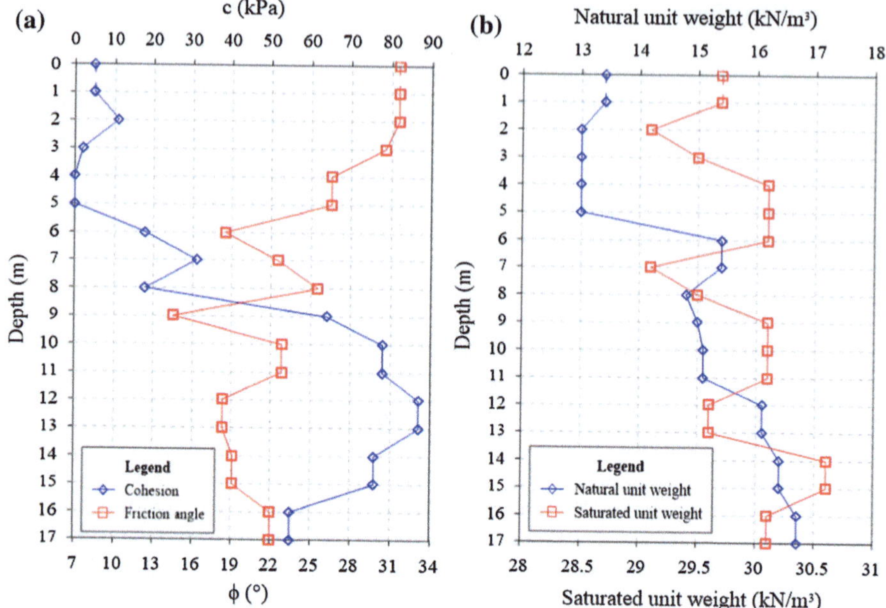

Fig. 2. (a) Mohr-Coulomb parameters (c and ϕ); (b) natural and saturated soil unit weight in the experimental field of Campinas.

3.1.2 Numerical Model

The numerical model is formed by 1419 quartic triangular elements, considering also an interface along the pile length in order to model the soil-structure behavior. In the calculation stage, the first step defines the initial stress conditions and the next phases assign the pile material properties to specific element clusters in order to simulate the pile construction.

3.1.3 Predicted Results

The stress and strain distributions corresponding to the last phase of pile loading considered in the numerical simulation are shown in Fig. 3(a)–(d), in which it can be seen that their maximum values are reached under the pile tip. The results presented in Fig. 4 indicate that the pile bearing capacity is 636 kN, in satisfactory agreement with the measured experimental value of approximately 660 kN reported by Albuquerque (2001). The theoretical estimate considering negative friction effects was 413 kN and the Décourt and Quaresma (1982) method was the closest to the loading test field, among the semi-empirical methods, with a bearing capacity of 483 kN. For the settlement control techniques, the Davisson (1972) and the Brazilian Standard NBR-6122 (1996) gave the best approximation to the field result, with estimates of 655 kN and 663 kN, respectively. Figure 5 shows the predictions obtained by six methodologies of load-settlement extrapolation, yielding estimated values from 530 kN to 700 kN.

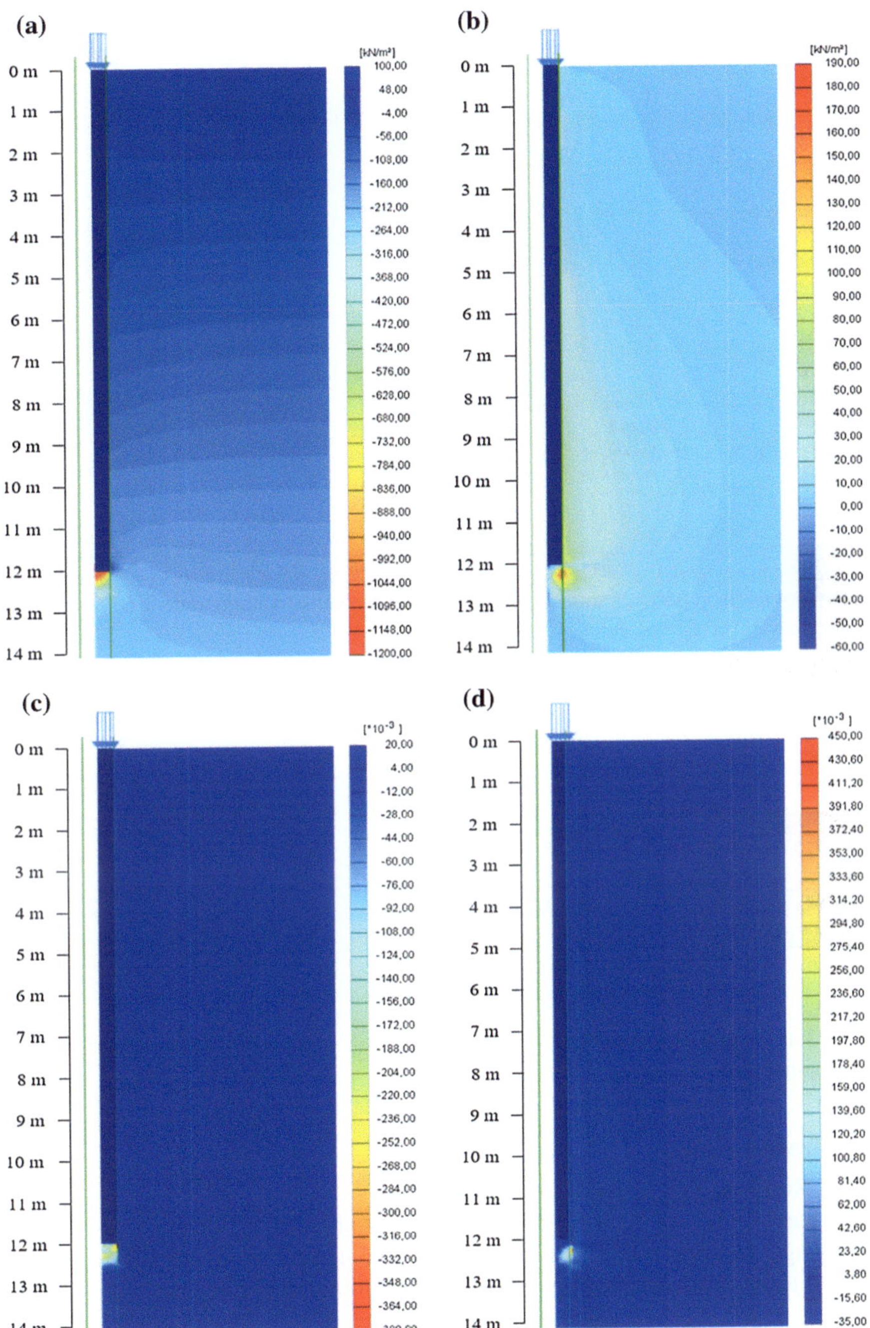

Fig. 3. Numerical results for Campinas's case: (a) total vertical stress σ_{yy}; (b) shear stress τ_{xy}; (c) vertical strain ε_{yy}; (d) shear strain γ_{xy}

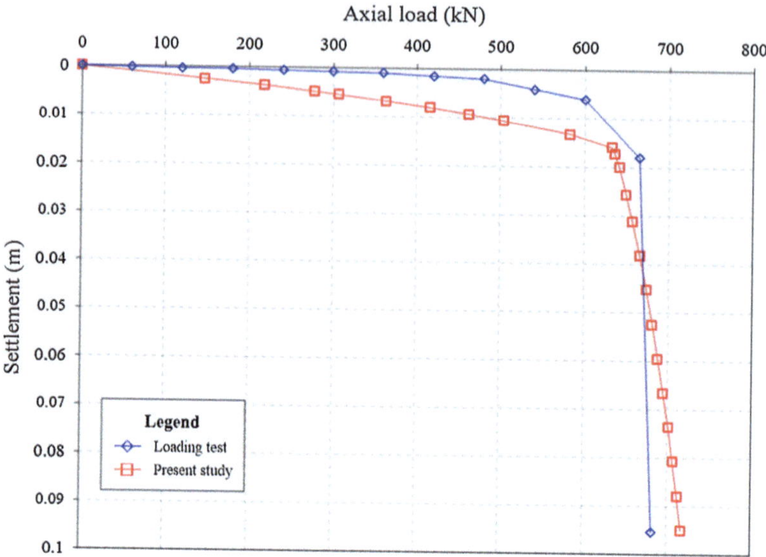

Fig. 4. Comparison between numerical and measured results (Albuquerque 2001) for Campinas case.

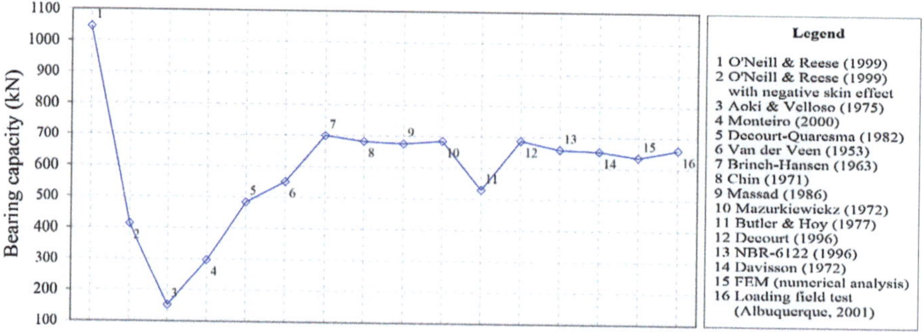

Fig. 5. Pile bearing capacity determined from analytical, semi-empirical and numerical methods for Campinas case.

3.2 Brasilia Case

3.2.1 Residual Soil Deposit and Parameters

In this case study five reinforced concrete piles (all with 0.3 m diameter, three with 7.5 m length and two with 8 m length) were tested in different months of the year. Two pile load tests are herein investigated, one bored in saturated soil condition and other in unsaturated (natural) soil condition, corresponding to high (November to March) and low (May to August) pluviometric precipitation levels, respectively. All tests were executed during several months with velocities given in Table 1. The experimental results reported by Mascarenha (2003) presented values of bearing capacity of 270 kN and 210 kN, under natural and saturated conditions, respectively. The soil profile and the corresponding SPT-N values along depth are shown in Fig. 6. Likewise, cohesion and friction angle values (Fig. 7) were obtained by Guimarães (2002) and Mascarenha (2003) for both soil foundation and soil-concrete interface. Shear strength parameters were obtained either in laboratory or correlated with SPT field test data.

Table 1. Summary of pile load tests in the experimental field of the University of Brasilia (Mascarenha 2003).

Load test	Pile	Beginning	End	Type	P_{max} (kN)	r_{max} (mm)	r_{final} (mm)
01	01	21/02/00	22/02/00	Slow	270	16.1	14.94
02	05	22/06/00	23/06/00	Slow	270	9.42	7.32
03	02	09/08/00	10/08/00	Slow	300	3.82	2.59
04	03	26/10/00	27/10/00	Slow	240	8.71	5.22
05	04	06/03/01	06/03/01	Slow	210	6.82	6.15
06	01	06/11/01	07/11/01	Slow	390	6.39	4.3
07	05	13/11/01	14/11/01	Slow	360	9.51	7.74
08	05	21/11/01	21/11/01	Slow	390	28.37	26.32
09	03	04/12/01	05/12/01	Slow	270	5.15	3.8
10	03	06/12/01	07/12/01	Slow	310	7.19	6.01
11	03	12/12/01	13/12/01	Slow	310	2.99	1.14
12	03	13/12/01	13/12/01	Fast	330	6.02	3.74
13	03	13/12/01	14/12/01	Fast	330	3.85	1.86
14	04	22/01/02	23/01/02	Slow	240	5.09	3.37
15	04	24/01/02	25/01/02	Slow	270	13	12.46
16	04	25/01/02	25/01/02	Slow	270	27.7	26.93
17	01	26/03/02	26/03/02	Slow	330	150.85	150.46
18	01	14/05/02	15/05/02	Slow	210	179.53	178.08
19	01	21/05/02	21/05/02	Slow	180	4.61	1.26
20	01	28/05/02	29/05/02	Slow	240	206.52	202.65
21	01	07/08/02	07/08/02	Slow	300	209.43	209.26

3.2.2 Brasilia Case Under Natural Soil Condition

From the finite element analysis with Plaxis 2D, the stress and strain distributions related to the last phase loading are illustrated in Fig. 8(a)–(d) and the predicted pile capacity of 268 kN agrees quite well with the measured load field test of 270 kN, according to Fig. 9. The analytical solution considering the negative friction skin effects yielded a value of 278 kN, which is also very close to the measured load in the field test. From the semi-empirical methodologies, the Décourt and Quaresma (1982) method gave the most conservative result, with an estimated bearing capacity of 240 kN. With regard to the results of settlement control techniques, the predictions are considered satisfactory with values of 264 kN (Davisson method) and 273 kN (Brazilian Standard NBR-6122). Among the methods based on the extrapolation of the load-settlement curves, the bearing capacity varied between 235 kN and 361 kN, as shown in Fig. 10.

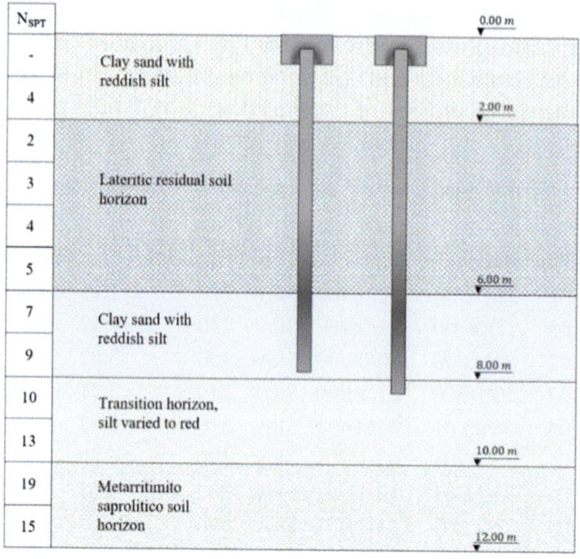

Fig. 6. Soil stratigraphy in the experimental field of the Brasilia case (modified from Mascarenha 2003).

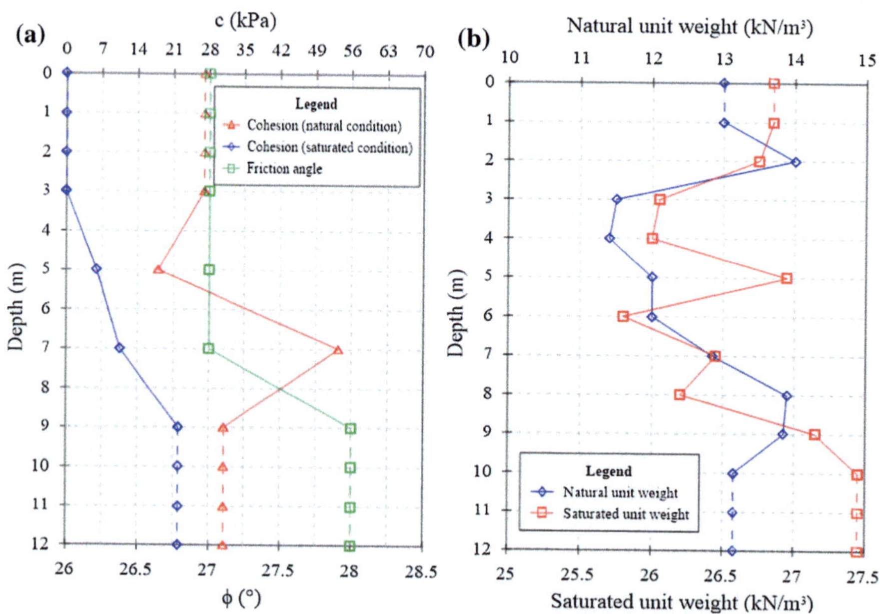

Fig. 7. (a) Mohr-Coulomb parameters (c and ϕ); (b) natural and saturated unit weight for soil in the experimental field of Brasilia (Mascarenha 2003).

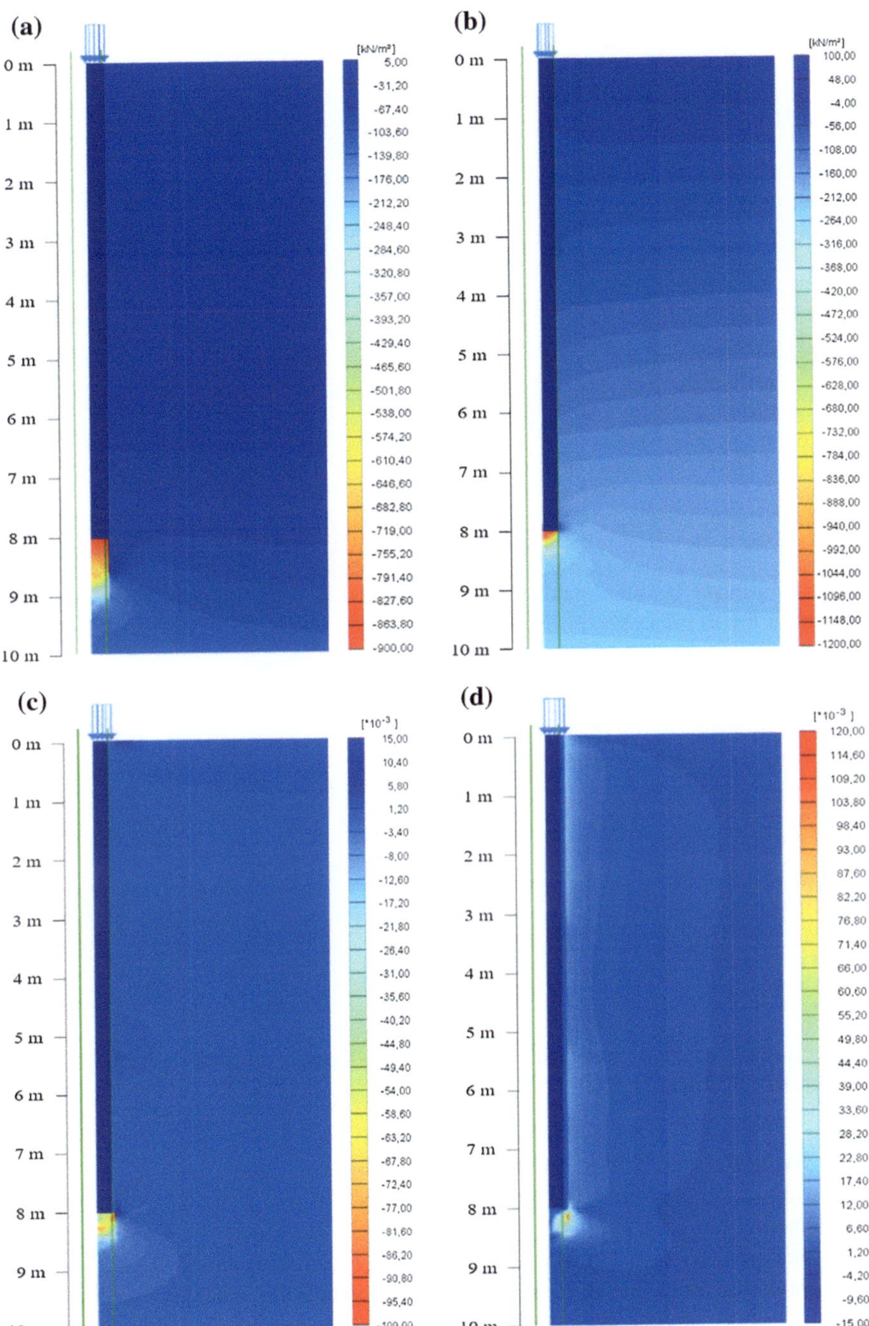

Fig. 8. Numerical results for Brasilia case under natural conditions: (a) vertical stress σ_{yy}; (b) shear stress τ_{xy}; (c) vertical strain ε_{yy}; (d) shear strain γ_{xy}.

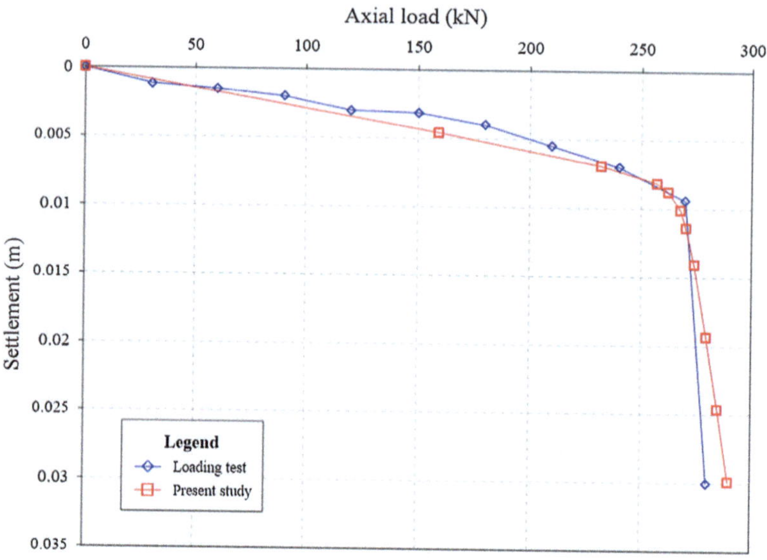

Fig. 9. Comparison between predicted and measured bearing capacity under natural soil condition for the Brasilia case.

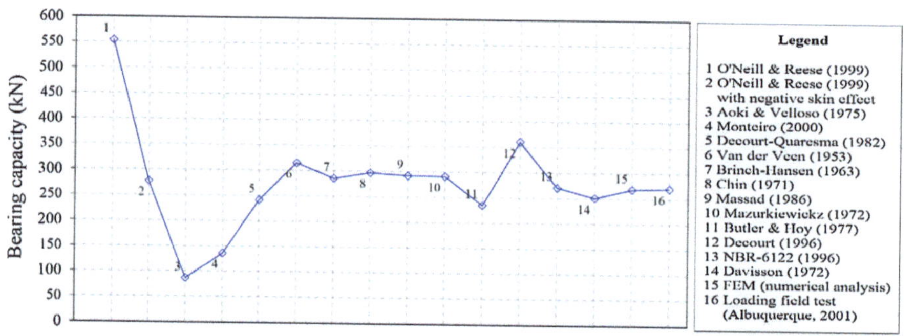

Fig. 10. Bearing capacity from analytical and semi-empirical methods for the Brasilia case under natural soil condition.

3.2.3 Brasilia Case Under Saturated Condition

The bearing capacity obtained from numerical analysis was 170 kN, while the ultimate load reached in the field test was reported as 210 kN. Both numerical and experimental load-settlement curves are compared in Fig. 11. From the analytical solution, taking into account the negative friction skin effects, a bearing capacity of 208 kN was estimated, while among the semi-empirical methodologies the Decourt-Quaresma method (1982) gave the closest result (241 kN). As for the Campinas case, the settlement control techniques resulted in satisfactory predictions of 210 kN, when applying the Davisson method (1972), and 211 kN, when using the NBR-6122

Brazilian Standard (1996). Regarding all the methods based on the extrapolation of the load-settlement curve, the results ranged between 188 kN to 239 kN, as indicated in Fig. 12.

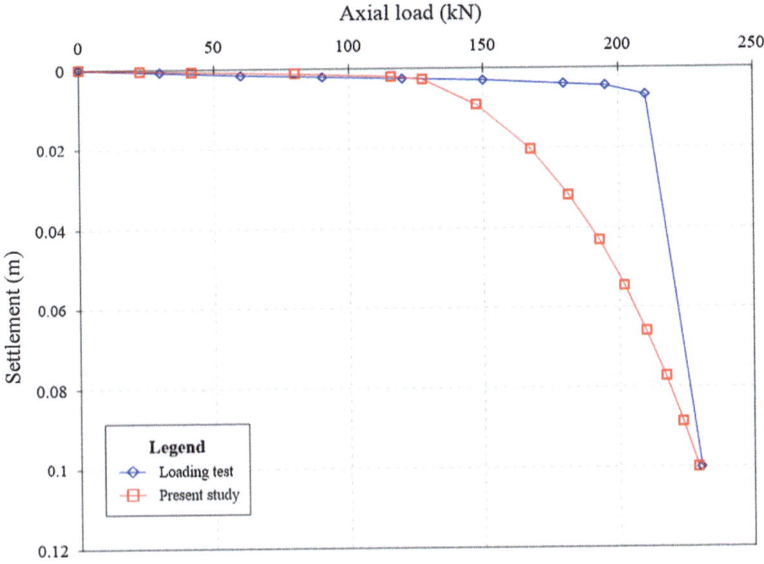

Fig. 11. Comparison between measured (Mascarenha 2003) and predicted numerical results for the Brasilia case under saturated condition.

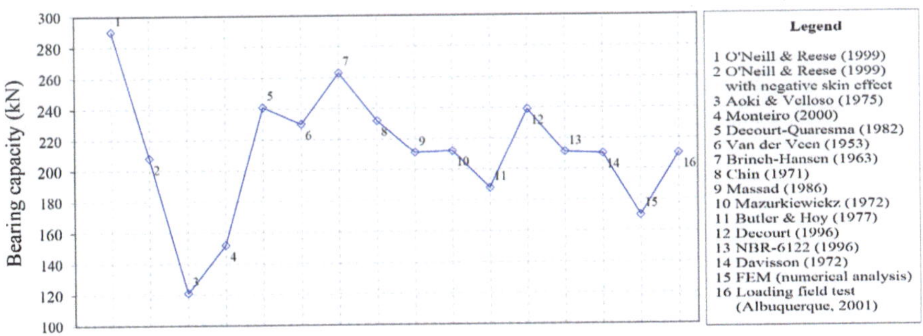

Fig. 12. Bearing capacity from analytical and semi-empirical methods for the Brasilia case under saturated condition.

4   Conclusion

Estimation of the bearing capacity of piles is an important subject in geotechnical engineering. The theoretical method that considered negative skin friction effects improved the bearing capacity results, although in the Campinas case, the predicted results were still away from the measured experimental values.

The semi-empirical bearing capacity methods were developed based on in-situ test data and are quite used in practical engineering applications. The results obtained for the three cases analyzed in this study, have shown that the Décourt and Quaresma (1982) method yielded the most accurate predictions with respect to the field tests.

About the two control of the settlement methods, the computed results were very close to the experimental data for all three cases herein investigated.

Concerning the different graphical extrapolation (load-settlement) methods, they showed reasonable results, although with a tendency to over predict the bearing capacity in much of methodologies for all the cases, with acceptable mean percentage errors of 8% (Campinas), 13% (Brasilia natural soil condition) and 17% (Brasilia saturated soil condition).

The numerical results computed through the finite element method gave good predictions for the Campinas and Brasilia cases (under natural soil condition). With respect to the Brasilia field test under saturated condition, the discrepancy observed in the computed result (170 kN), when compared to the experimental value (210 kN), may be due to the type of analysis considered in the numerical simulation (drained analysis) while the field test, carried out in just one day, probably needed a further analysis under undrained condition.

Acknowledgments. The first author would like to thank professor Celso Romanel and the Civil and Environmental Engineering Department at Pontifical University Catholic of Rio de Janeiro for supporting this research.

References

Albuquerque, P.J.R.: Estacas Escavadas, Hélice Contínua e Ômega: Estudo do Comportamento à Compressão em Solo Residual de Diabásio, através de Provas de Carga Instrumentadas em Profundidade, Tese de Doutorado, Escola Politécnica da USP, 260p (2001)

Aoki, N., Velloso, D.D.A.: An approximate method to estimate the bearing capacity of piles. In: Proceedings of 5th Pan-American Conference of Soil Mechanics and Foundation Engineering, vol. 1, pp. 367–376. International Society of Soil Mechanics and Geotechnical Engineering, Buenos Aires (1975)

Brinch-Hansen, J.: Hyperbolic stress-strain response: cohesive soils. ASCE JSMFD **89**(SM4), 241–242 (1963)

Butler, H.D., Hoy, H.E.: The Texas quick load test method for foundation load testing–users manual. In: FHWA IP-77.8, FHWA Implementation Division, Washington, DC (1977)

Chin, F.K.: Pile tests: Arkansas river project. JSMFD ASCE **97**(SM7), 930–932 (1971)

Davisson, M.T.: High capacity piles. In: Proceedings of Soil Mechanics Lecture Series on Innovations in Foundation Construction, pp. 81–112 (1972)

Décourt, L., Quaresma, A.R.: Como calcular (rapidamente) a capacidade de carga limite de uma estaca. A Construção, São Paulo (1982)

Décourt, L.: A ruptura de Fundações Avaliada com Base no Conceito de Rigidez. IIISEFE São Paulo **1,** 215–224 (1996)

Guimarães, R.C.: Análise das propriedades e comportamento de um perfil de solo laterítico aplicada ao estudo do desempenho de estacas escavadas. Dissertação de Mestrado, Publicação G.DM-091A/02, Departamento de Engenharia Civil e Ambiental, 183p, Universidade de Brasilia, Brasilia, DF (2002)

Mascarenha, M.D.A.: Influência do recarregamento e da sucção na capacidade de carga de estacas escavadas em solos porosos colapsíveis. Dissertação de Mestrado, Publicação no G. DM-098A/03, Departamento de Engenharia Civil e Ambiental, Universidade de Brasília, Brasília, DF (2003)

Massad, F.: Notes on the interpretation of failure load from routine pile load tests. Revista Solos e Rochas São Paulo 9(1), 33–36 (1986)

Mazurkiewicz, B.K.: Test loading of piles according to Polish regulations. R Swed Acad Eng Sci Comm Pile Res (35), 20 (1972)

NBR 6122: Projeto e execução de fundações. ABNT, Rio de Janeiro (1996)

O'Neil, M.W., Reese, L.C.: Drilled shafts: construction procedures and design methods (No. FHWA-IF-99-025) (1999)

Tomlinson, M.J.: The adhesion of piles driven in clay soils. In: Proceedings of the 4th International Conference on Soil Mechanics and Foundation Engineering, vol. 2, pp. 66–71 (1957)

Tomlinson, M.J.: Some effects of pile driving on skin friction. In: Behavior of Piles, pp. 107–114. Thomas Telford Publishing (1971)

Van der Veen, C.: The bearing capacity of a pile. In: 3rd International Conference on Soil Mechanics and Foundation Engineering, Zurich, vol. 2, pp. 84–90 (1953)

Utilizing Secant Pile Walls as Retaining Structures and Bridge Abutments

M. I. Y. Elzain[1(✉)] and M. Dafalla[2]

[1] School of Civil Engineering, Sudan University of Science and Technology, Khartoum, Sudan
elbuny@gmail.com
[2] Bugshan Research Chair in Expansive Soils, Civil Engineering, King Saud University, Riyadh 11421, Saudi Arabia
mdafalla@ksu.edu.sa

Abstract. This study presents a design solution for a highway underpass beneath a busy main road in Hail city, Saudi Arabia. The traffic intensity within the project area is very high and some buildings are located very close to the highway. An initial design suggested by the ministry of transportation (MOT, Saudi Arabia) included a sheet pile system and retaining wall abutments. The geotechnical investigation conducted for the site revealed that layers of sandy soils will require the use of sheet piles to prevent loss of lateral support or damage to the foundations of the adjacent buildings and service roads. This study suggested the use of dual-purpose secant piles walls capable of supporting the subsurface soil and provides fewer disturbances to traffic flow during the construction phase. The spread footing and 13.5 m retaining wall design suggested as a first solution was found more expensive and takes longer period of construction. Details of the secant pile walls procedure are given and compared to the conventional retaining wall and sheet pile solution. Cost comparison and elements of choice were highlighted and provided in more details.

1 Introduction

Retaining structures are widely used to give lateral support for earth or other non-stable material in order to enable successful facility construction. These walls are used in different applications including protection of existing structure, highway embankments, road widening, subways and underpass structures as well as many applications that need temporary or permanent support (Dafalla et al. 2016). Active, passive or at-rest earth pressure is expected to act on typical walls. The work presented in this paper calls for the use of secant pile walls as an economical alternative solution when the retaining structure is supporting a deep excavation that requires lateral support in the form of sheet piles. The secant pile system can also be used when limited construction space is available and especially appropriate in congested urban areas where noise and the effect on adjacent properties were to be kept to minimum (Municipality of Riyadh 2014).

The sites for two interchanges at a ring road in Hail city, Saudi Arabia known as Algalam Interchange were investigated for the geotechnical and engineering properties (Figs. 1 and 2).

© Springer International Publishing AG, part of Springer Nature 2019
W. Frikha et al. (eds.), *New Developments in Soil Characterization and Soil Stability*,
Sustainable Civil Infrastructures, https://doi.org/10.1007/978-3-319-95756-2_7

Table 1. Typical soil profile and silty sand composition of an ideal borehole

| Sample reference | | S.P.T (N) | W.C % | Atterberg limits | | | Grain size distribution | | | | USCS classification | AA SHTO classification |
Number	Depth (m)			L.L %	PL %	P.I %	% Gravel	% Sand	% Silt	% Clay		
B-1-1	0.1	36	1.5	–	–	NP	16.4	67.7	15.9		SM	A-1-b
B-1-2	1.5	44	1.7									
B-1-3	3.0	47	2.1	–	–	NP	10.9	75.7	13.4		SM	A-1-b
B-1-4	4.5	64	2.6									
B-1-5	6.0	71	2.8									
B-1-6	7.5	87	2.9									
B-1-7	9.0	89	3.1									
B-1-8	10.5	100	3.5									
B-1-9	12.0	100	4.6									
B-1-10	13.5	100	6.3	–	–	NP	11.6	74.7	13.8		SM	A-2-4(0)
B-1-11	15.0	100	8.1									
B-1-12	16.5	100	9.6									
B-1-13	18.0	100	12.1									
B-1-14	19.5	100	12.6									
B-1-15	21.0	100	12.9									
B-1-16	22.5	100	12.8	–	–	NP	7.6	75.5	16.9		SM	A-2-4(0)
B-1-17	24.5	100	13.1									

Table 2. Cost calculations for secant piles as bridge abutment

Description	Unit	Quantity	Rate (US $)	(US $) Total
Pile drilling (1.5 and 1.0 m dia 12 m under the bridge and varying depth as in retaining wall.)	Ml	5900	100	590,000
Concrete of un-reinforced primary piles	M3	5000	100	500,000
Reinforced concrete secondary piles including pile testing	M3	5000	250	1,075,000
Reinforcing steel including fabrication and erection.	tones	2000	1000	2,000,000
Total costs				4,165,000
Benefits gain of 11 months' time saving				−1,015,000
Net costs of the used of secant piles pile construction				**3,150,000**

Table 3. Cost Calculations for conventional abutment

Description	Unit	Quantity	Rate (US $)	(US $) Total
Excavation for abutment and retaing walls	M3	43000	5	215,000
Plane concrete (0.1 m thick) under foundation	M3	620	100	62,000
Reinforced concrete for abutment & retaining walls	M3	15000	215	3,225,000
Reinforcing steel	tones	3000	1000	3,000,000
Water proofing	M2	47000	5	235,000
Back filling of foundation	M3	34000	5	170,000
Cost of 5 km temporary road	job			2,920,000
Ing time in construction of 11 month	1			1,015,000
Total costs				**10,842,000**

The subsurface soil information provided by the twelve (12) borings drilled to depths of 20–25 m revealed that the subsurface formations at the site consist basically of three layers extending from the ground surface to the maximum depths investigated.

The site was generally overlain with fill material of 1.5–2 m depth covered with an asphalt pavement. Below the fill a silty sand material of variable densities extend to the bottom of investigated boreholes. Groundwater table was encountered at a depth of −17.00 M to −18.00 M below ground level (Table 1).

Such typical investigation will avoid future cracks or problems for the retaining wall (Al Fouzan and Dafalla 2014).

For the lateral earth pressure behind the abutments, the following values of density (γ) of granular fill, active earth pressure coefficient (k_a) and coefficient of friction (f) between footing and soil were recommended.

$$\gamma = 18.0 \ kN/m^3$$
$$K_a = 0.333$$
$$f = 0.40$$

Lateral loads on the retaining abutment walls were computed based on these parameters. The angle of internal friction (Φ) for the granular soil of a density (γ) equals to 18.0 kN/m^3 can be taken as $\Phi = 33°$ (Das 2010).

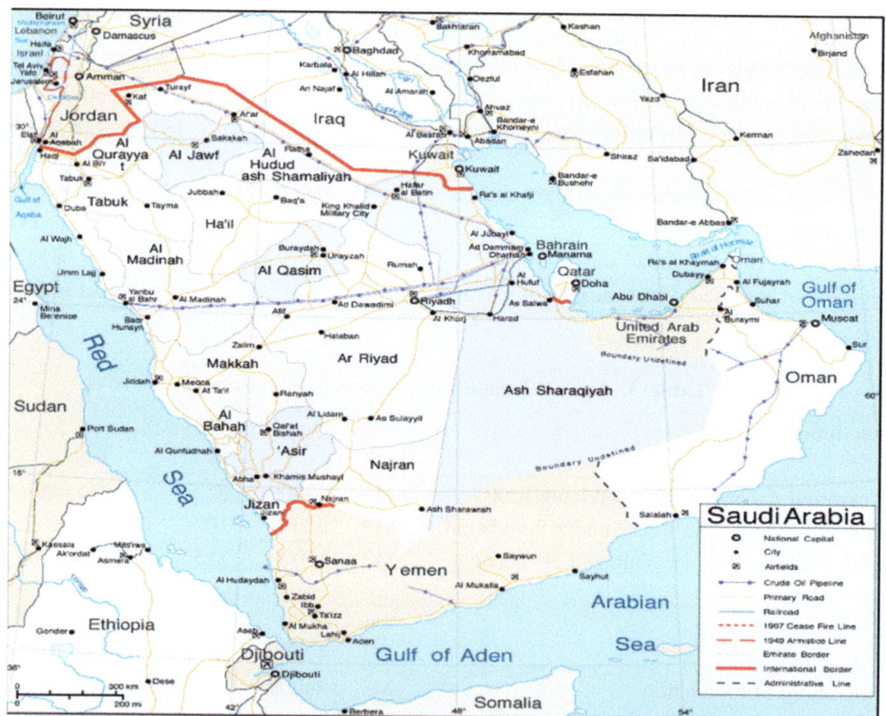

Fig. 1. The site location- Hail city in Saudi Arabia.

Fig. 2. A satellite view of the project site.

2   Site Overview

A proposed dualization underpass bridge for Ring Road Project in Hail city in the North of Saudi Arabia is considered for this study. The proposed underpass bridge consists of two carriageways each 21 m wide and 96 m long supported on central piers and two abutments. The retaining walls extend over a length of 800 m with a variable height according to underpass profile. The main road passes through a busiest commercial and residential area with a service road consisting of three lanes. The service road is three meter away from the existing buildings. Where the tracks cross Suad bin Abdelmuhsin Street on the heart and central area of Hail city, with a high volume of traffic, Fig. 3 presents the proposed project geometry. The layout of local streets and businesses effectively precluded construction of such conventional retaining wall. Structural designer recommended the sheet pile to support lateral loads, the main and service roads from both directions should be closed, the traffic converted via temporary roads of two lanes each having lengths of five kilometer. The proposed abutment foundation is 11 m in width, Figs. 4 and 5 presents a cross section of the proposed bridge.

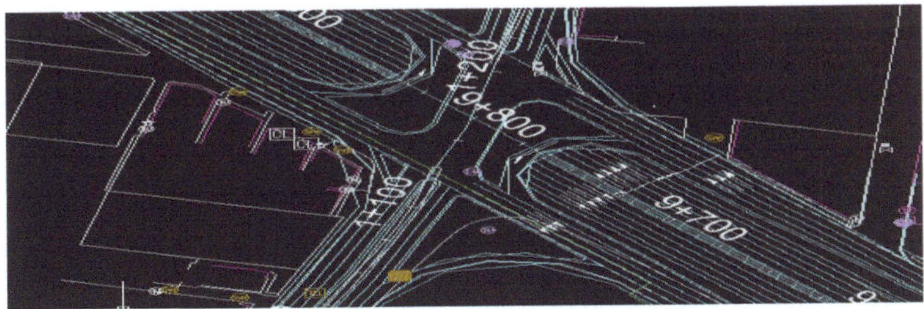

Fig. 3. Proposed geometrical plan layout.

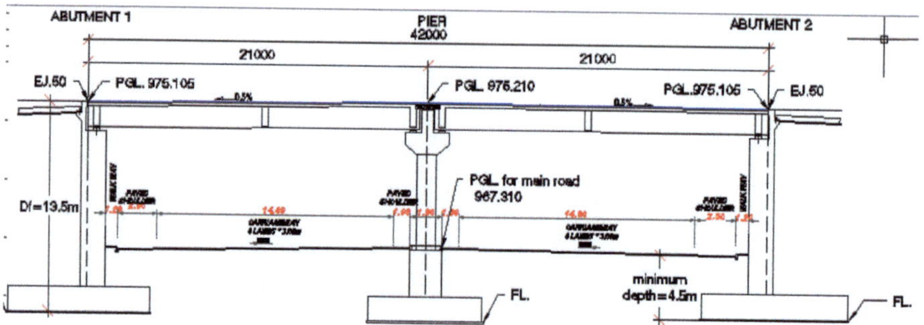

Fig. 4. Typical cross section of the bridge.

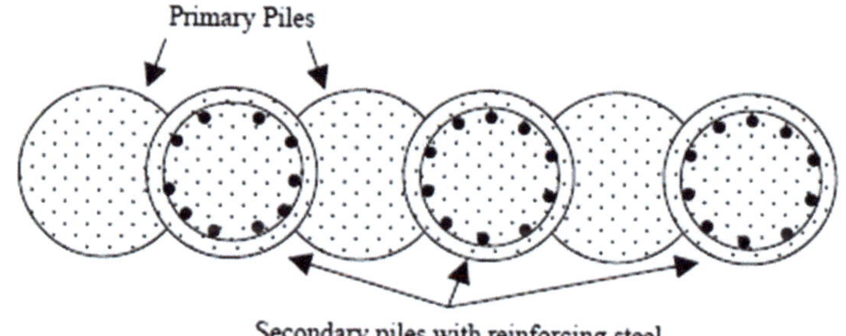

Fig. 5. Cross-section of secant piles wall.

3 Secant Pile Suggestion

The alternative design suggested for the underpass was proposed as a cut-and-cover structure, with secant pile walls that act as soil retaining walls and a support structure with 192 piles each of 1.5 m diameter carrying vertical and lateral loads and work as bridge abutments, 800 piles of one meter diameter as the retaining walls carrying lateral loads. Finally the superstructure portion, or cover, is placed on the support structure. Secant pile walls are constructed by first drilling a line of un-reinforced primary piles with a clear spaced approximately half a diameter apart. The concrete in these piles is designed to have a relatively low strength, so that it can be easy to drill. The reinforced secondary piles with the same diameter are drilled between the primary piles while the concrete of the primary piles is still young. The secondary piles cut into the primary piles to form an interlocked wall.

4 Evaluation of Alternatives

This study presents an evaluation of the project alternatives according their functions using project management tools such as value engineering process or cost benefit process. The cost benefit analysis is a term that refers to the process that helps to appraise, or assess the case of a project proposal and approach in order to make an economic decision. The process evolves whether explicitly or implicitly, weighing the total expected costs or more actions in order to choose the best or more profitable option. The formal process is often referred to as (CBA) cost-benefit analysis or (BCA) benefit- cost analysis. Benefits and costs are often expressed in many terms and adjusted for the time value of money so that all flows of benefits and project costs over time and the common unit of measure is (money).

The planned construction schedule indicates the use of secant piles construction will save eleven months when compared to the conventional retaining wall option (Tables 2 and 3). The gain from finishing the project eleven month earlier than estimated period of four years was calculated using cost benefit analysis tools and found to be US $ 1,015,000.

5 Conclusions

This study presented a creative solution of low cost that can benefit project owners in term of money and time. The secant piles wall proposal is suitable as an alternative for deep retaining walls within granular silty sand soil where expensive support of lateral sides of excavation is required. Project management tools can be utilized to assess the gain of using such type of foundation and support for typical highway underpasses.

Acknowledgments. The authors appreciate the support of King Saud University, Deanship of Research Chairs.

References

Das, B.M.: Principles of Foundation Engineering, 7th edn. Christopher M. Shortt, Publisher Global Engineering (2010)

Municipality of Riyadh: Design of abutment and retaining walls for Prince Turki intersection with Musa Bin Nasir and Prince Mohammed bin Abdulaziz roads underpass. Riyadh, Shibh–Aljazira (2014)

Dafalla, M.A., Al-Shamrani, M.A., Al Subaie, F.S., AlFouzan, S.K., Charif, A.: Comparisons of two approaches for the design of retaining walls supported on drilled piles. Geotechn. Struct. Eng. Congr. **2016**, 2035–2042 (2016)

Al Fouzan, F., Dafalla, M.A.: Study of cracks and fissures phenomenon in Central Saudi Arabia by applying geotechnical and geophysical techniques. Arab. J. Geosci. 7(3), 1157–1164 (2014)

Stabilization on 1000 kv UHV Transmission Tower Foundation Influenced by Underground Coal Mining in Mountainous Areas

Bo Liu[1,2(✉)], Yongjun Ma[1], Bo Cao[3], and Weijie Xu[4]

[1] School of Mechanics and Civil Engineering,
China University of Mining and Technology, Beijing, China
liub@cumtb.edu.cn

[2] State Key Laboratory of Geomechanics and Deep Underground Engineering,
Beijing, China

[3] China Energy Engineering Group Guangdong Electric Power Design Institute
Co., Ltd, Guangzhou, China

[4] The Architectural Design and Research Institute of Henan Province Co., Ltd,
Zhengzhou, China

Abstract. This paper presents a case study of stabilization of a transmission tower foundation influenced by underground coal mining on mountains. The tower foundation is built on land that is going to be mined, and it belongs to a 1000 kV UHV (ultra high voltage) transmission line in Shanxi province. Considering the synergistic effects between rock and transmission tower foundation, the transmission tower foundation deformation and the potential dangerous situations were calculated by using FLAC^{3D} code numerical simulation. The comparison of results by numerical and analytical calculation was conducted as well. According to simulation results, the tower foundation deformation is dominated by the effect of coal mining rather than rock slope sliding. The potential moving track of tower foundation surface points is incomplete "C" shape in adverse slope situation. The modified reinforced concrete slab foundation combined with protection panel and length-adjustable anchor bolts is recommended in this paper. The UHV tower foundation construction followed the recommended method was completed in 2015. Further field monitoring data show that the modified tower foundation can be suggested to guarantee the safety of transmission tower during coal seam mining.

1 Introduction

It is normal for the overhead high-voltage transmission lines to run across infection zones of coal mining. The stability of transmission tower will be influenced as it is sensitive to surface deformation (Qianjin et al. 2012; Fengli et al. 2013). A 1000 kV UHV transmission line runs across Shanxi Province where there are lots of mountains and numerous mined-out areas. Longwall mining method along strike with natural caving is used in this mining areas and it causes large-scale surface deformation (Khanal et al. 2014; Li et al. 2016; Likar et al. 2012). Generally, reserving security coal pillars and proper protection areas are the usual protective methods to ensure the safety

W. Frikha et al. (eds.), *New Developments in Soil Characterization and Soil Stability*,
Sustainable Civil Infrastructures, https://doi.org/10.1007/978-3-319-95756-2_8

of the upper buildings (Kratzsch 1983; Ping et al. 2014). But this causes lots of coal resources waste. To avoid this situation, the current design codes in electric system suggest that enough tower foundation deformations should be reserved. Guanglin et al. (2009) introduced a kind of composite foundation which is used to guarantee the stability of the transmission tower in mining subsidence area. Fengli et al. (2013) analyzed the stability of transmission tower influenced by underground coal mining using ANSYS software. Yang established transmission tower model and applied different deformations on tower foundation to simulate the effect of coal seam mining. But there are few relevant researches about transmission towers on mountains. It is important to study the stabilization of UHV transmission tower foundation influenced by underground coal mining on mountains, the synergistic effect between rock and tower foundation should also be considered.

In this paper, a tower foundation which belongs to the 1000 kV UHV transmission line is selected. The tower foundation locates on an adverse rock slope where the land is going to be mined. The modified reinforced concrete slab foundation combined with protection panel and length-adjustable anchor bolts is used in this UHV tower foundation as shown in Figs. 1 and 2. Parameters of this UHV tower foundation are shown in Table 1.

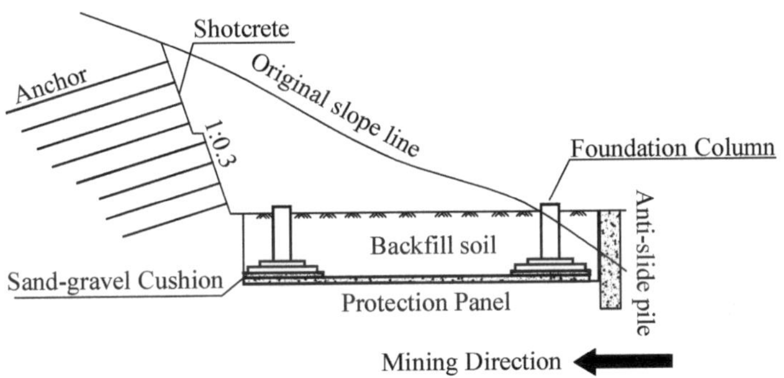

Fig. 1. Design of UHV transmission tower foundation

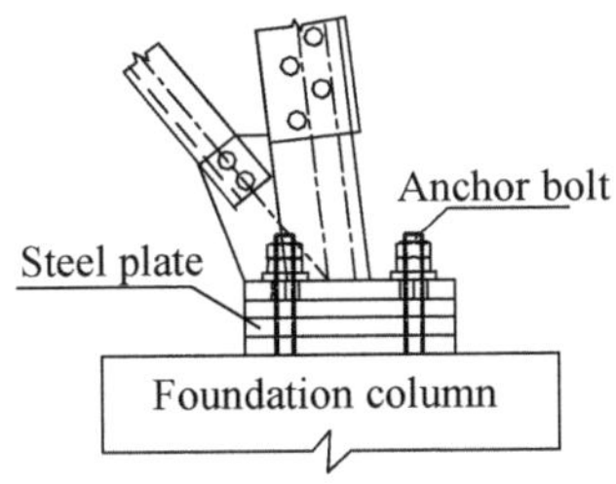

Fig. 2. Length-adjustable anchor bolts

Table 1. Parameters of transmission tower foundation

Items	Parameter	Items	Parameter
Size of slab	31 m × 31 m	Slope height	25.6 m
Cross section of anti-slide pile	1.2 m × 1.7 m	Anchor diameter	$\varnothing$25 mm
Thickness of retaining plank	0.3 m	Anchor length	10 m
Shotcrete strength	C20	Anchor spacing	2 m × 2 m
Slope angle	1:0.5	Anchor angle	25°

2 Modeling Analysis Based on Synergistic Effect Between Rock and Tower Foundation

The Fast Lagrangian Analysis of Continua (FLAC^{3D}) numerical model is shown in Fig. 3a. Model sizes are 600 m (X), 1000 m (Y) and 600 m (Z). Coal seam thickness is 6.0 m and the average buried depth is 450 m. In order to consider the synergistic effect between tower foundation and rock, all the items such as tower foundation, protection panel, anti-slide piles, retaining plank, high slope, shotcrete and anchors, are modeled according to the design scheme.

All the rocks are assumed as ideal elastoplastic body obeying Mohr-Coulomb yielding criteria. Physical and mechanical parameters of rocks are shown in Table 2. The loads from upper tower structures are distributed on the top of four foundation columns equivalently. The design values of additional stresses on the top of foundation columns are shown in Table 3.

Longwall coal mining method along strike with natural caving is used in this area. The arrangement of working faces is shown in Fig. 3a. The width of each working face is 180 m; the width of each coal pillar between adjacent working faces is 30 m. Working face 1 is mined completely first followed by working face 2 and 3. The mining height is 6.0 m. As shown in Fig. 3b, on the top of foundation columns, four monitoring points marked with A ~ D were set up to monitor the deformations of tower foundation during the whole process of coal mining.

3 Results and Analysis

3.1 Tower Foundation Settlement

Settlement curves of tower foundation are shown in Fig. 4. It is shown that the tower foundation keeps sinking and ends up with 3.24 m. In mountainous areas, the settlement consists of two components caused by coal mining and rock slope sliding. For each working face, when mining towards the tower foundation, both of the two components will increase faster and faster and they will reach the maximum rates when the working face is under tower foundation. After that, the two components will still increase, but the increase rate is slowdown. At last the tower foundation stops moving. In the whole process, the moving directions of the two components are the same.

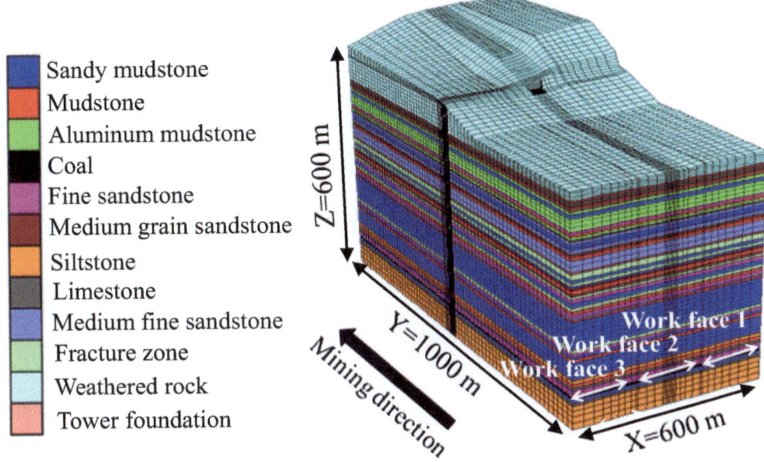

(a) Integrated 3-D numerical model

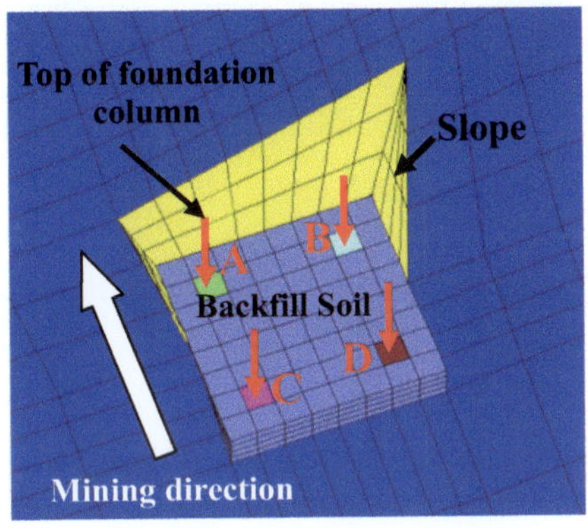

(b) UHV tower foundation

Fig. 3. Integrated 3-D numerical model of underground coal mining and UHV tower foundation on mountains

The settlements caused by working face 1–3 are 0.67 m, 1.78 m and 0.79 m respectively. The working face 2 plays the most important role, accounting for 55% of the whole settlement.

3.2 Horizontal Displacement of Tower Foundation

The horizontal displacement curves are shown in Fig. 5.

For tendency displacement, when mining coal working face 1, the tower foundation moves towards positive X direction and the final displacement is 0.4 m. When mining coal working face 2, the tower foundation moves towards negative X direction. Thus, the total tendency displacement decreases and the final value is 0.27 m towards

Table 2. Physical and mechanical parameters of rocks

Lithology	Tensile strength/MPa	Cohesion/MPa	Friction/ (°)	Shear modulus/MPa	Bulk modulus/MPa
Fine sandstone	2.78	2.60	38	7885	17083
Medium fine sandstone	2.17	2.00	34	7311	17870
Medium grain sandstone	1.94	1.90	33	5733	17200
Sandy mudstone	1.32	1.32	30	5096	14225
Mudstone	0.90	0.98	31	5012	13992
Siltstone	2.78	2.56	35	10779	15655
Limestone	5.51	5.51	48	11047	22619
Fracture zone	0.02	0.03	20	492	1886
Aluminum mudstone	1.13	1.13	31	5134	14331
Coal	0.63	0.79	25	4529	17361

Table 3. Design values of additional stresses of foundation (kN)

	Uplift force			Downward force		
	T	Fx	Fy	N	Fx	Fy
Design value	3634	628	585	5268	883	788

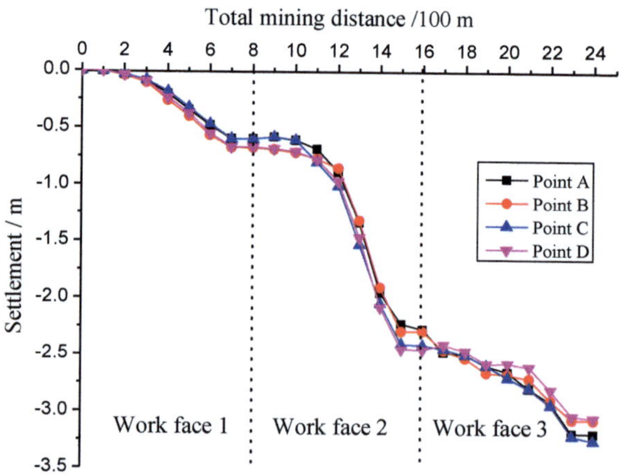

Fig. 4. Settlement curves of UHV tower foundation

positive X direction. At last, mine coal working face 3. The tower foundation moves towards negative X direction and the ultimate displacement is 0.3 m towards negative X direction.

The strike displacement consists of two components that are caused by coal mining and rock slope sliding. For each working face, when mining towards the tower foundation, both of the two components will increase faster and faster towards the negative Y direction, they will reach the maximum rates when the working face is under tower foundation. The two components are in the same direction. After that, the increase rate of the two components is slowdown and their directions are the opposite. On the whole, when mining each working face, the strike displacement increases first and then decreases, but the ultimate displacement value is larger than the original value. This indicates that the deformation of tower foundation is dominated by coal mining rather than rock slope sliding. The potential moving track of tower foundation surface points is incomplete "C" shape in adverse slope situation. This is different from the situations in plain areas and soil slopes (Wenxiu et al. 2010; Zhaoping et al. 2016). The maximum strike displacements caused by working face 1 to 3 are 0.24 m, 0.92 m and 0.72 m respectively, and the ultimate value is 0.56 m.

3.3 Tilt of Tower Foundation

There is a directly proportional relationship between tilt and horizontal displacement of tower foundation, as shown in Figs. 5 and 6.

The trend of tendency tilt is the same as horizontal displacement. When mining coal working face 1, the tower foundation tilts towards positive X direction, the final value is 3.6 mm/m. When mining coal working face 2, the tilt value decreases. The final tilt towards positive X direction still and the value is 1.52 mm/m. When mining coal working face 3, the tilt value decreases further and at last the tilt is towards negative X direction. The ultimate value is 2.5 mm/m.

The trend of strike tilt is the same as horizontal displacement. When mining each working face, the tower foundation tilts towards negative Y direction. The value

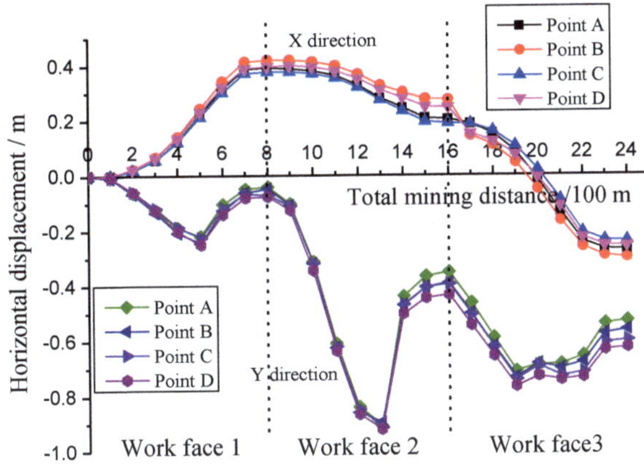

Fig. 5. Horizontal displacement curves of UHV tower foundation

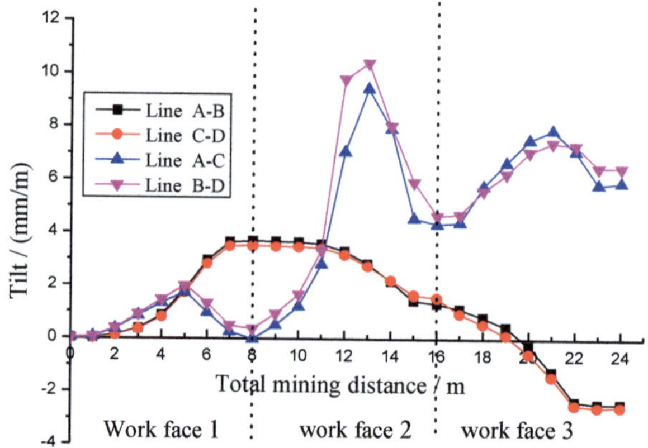

Fig. 6. Tilt curves of UHV tower foundation

increases first and then decreases, but the final value is larger than original value presenting an incomplete "C" shape. The maximum tilt values caused by each working face are 1.9 mm/m, 10.4 mm/m, 7.8 mm/m respectively, and the ultimate value is 6.4 mm/m.

4   Analytical Calculation of Surface Deformation

Usually, the working face is regarded as a semi-infinite mining condition, and probability integral method is often used to predict the surface deformation in electric system. It should be noted that this analysis ignores the coal pillar functions. In this paper, the probability integral method based on superposition principle is applied to predict the surface deformation, which assumes that the surface deformation conforms to superposition principle. The superposition principle is shown in Fig. 7.

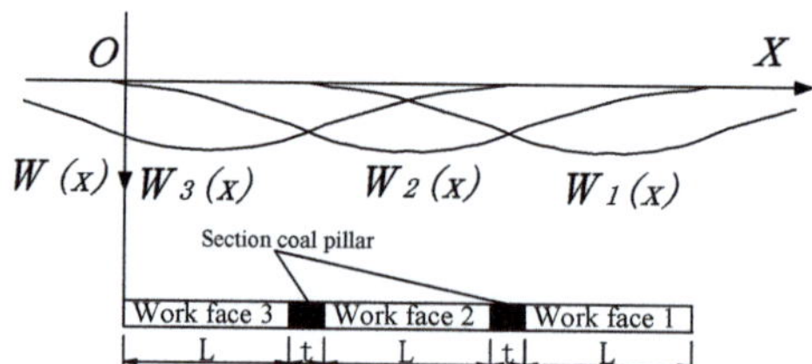

Fig. 7. Superposition principle of surface deformation

4.1   Modified Formulas for Surface Deformation

The surface keeps sinking during the whole process of coal mining. Based on the deformation prediction formulas for any points in subsidence basin (Guoqing et al. 1994), the modified formulas are shown below.

$$W(x,y) = \frac{1}{W_3}W_3^0(x)W_3^0(y) + \frac{1}{W_2}W_2^0(x-L-t)W_2^0(y) + \frac{1}{W_1}W_1^0(x-2L-2t)W_1^0(y) \qquad (1)$$

$$W_i = m_i q_i \cos \alpha_i \qquad (2)$$

Where, W_i is the maximum surface settlement value when working face i is mined in semi-infinite mining condition; $W_i^0(x)$ is the surface settlement value of points whose tendency coordinate is x when working face i is mined in strike full mining condition; $W_i^0(y)$ is the surface settlement value whose strike coordinates is y when working face i is mined in tendency full mining condition; m_i is the mining height of working face i; q_i is the subsidence coefficient of working face i; α_i is the coal seam dip of working face i.

The tower foundation is located in the middle of working face 2. The maximum horizontal displacement and tilt of tower foundation are towards strike direction. The deformation caused by working face 1 and 3 is very small, so the strike deformation caused by working face 2 is considered, which can be simplified as tendency finite mining while strike full mining condition.

$$U^0(y) = C_{xm}(U(y) - U(y - L)) \qquad (3)$$

$$i^0(y) = C_{xm}(i(y) - i(y - L)) \qquad (4)$$

$$C_{xm} = \frac{W_{mx}^0}{W^0} \qquad (5)$$

Where, $U^0(y)$, $i^0(y)$ are the horizontal displacement and tilt values along strike main section when it is tendency finite mining condition; $U(y)$, $i(y)$ are the horizontal displacement and tilt values along strike main section when it is semi-infinite mining condition; W_{mx}^0 is the maximum settlement value along tendency main section when it is tendency finite mining condition; W^0 is the maximum settlement value when it is semi-infinite mining condition.

4.2 Comparison Between Analytical and Numerical Results

According to field monitoring results, $q = 0.82$, $\tan \beta = 2$, $b = 0.25$ and the recovery ratio is 85%. Based on Formula (1)–(5), the comparison between analytical and numerical calculation results is shown in Table 4. The maximum analytical deformations are 3.14 m, 0.85 m, 9.36 mm/m for settlement, horizontal displacement and tilt separately. Then in FLAC^{3D} case, the corresponding values are 3.24 m, 0.92 m and 10.4 mm/m. The results in analytical solution have a good agreement with the numerical calculations.

Table 4. Comparison between analytical and numerical calculation results

	Settlement/m	Horizontal displacement/m	Tilt/(mm/m)
Analytical calculation	3.14	0.85	9.36
Numerical calculation	3.24	0.92	10.4

5 Engineering Practice

5.1 UHV Tower Foundation Scheme

Protection panel can suffer a large deformation. According to practical experience, when using protection panel, the transmission tower incline generally with no significant changes in root span. Based on the investigation and calculation results, the modified reinforced concrete slab foundation combined with protection panel and length-adjustable anchor bolts is used in this UHV transmission line project.

Generally, the uniform settlement will not destroy the transmission tower structure, but enough preset deformation should be considered in advance for the transmission line. Lengthen the preset length of the exposed section of anchor bolts so the steel plates can be installed to rectify the tilt of the transmission tower. Lay sand-gravel cushion between protection panel and reinforced concrete slab foundation. The sand-gravel cushion reduces the friction between protection panel and slab foundation, so it makes thrusting conveniently when the tower foundation has large horizontal displacement and need to be rectified.

5.2 Construction and Field Monitoring

According to numerical and analytical results as well as data investigation, enough preset deformations are considered in the design of UHV tower foundation. The recommended UHV tower foundation construction is completed in 2015. The construction of the protection panel in the field is shown in Fig. 8. In the field, the deformations of tower foundation were monitored based on the monitor points set up on the top of foundation columns. Up to now, working face 1 has been mined for about 400 m, the comparison between numerical calculation and field monitoring results is shown in Fig. 9. The settlement of tower foundation surface is about 18.65 cm and the tower foundation is secure.

Fig. 8. Construction of the UHV tower foundation on mountains; Protection panel was being pouring

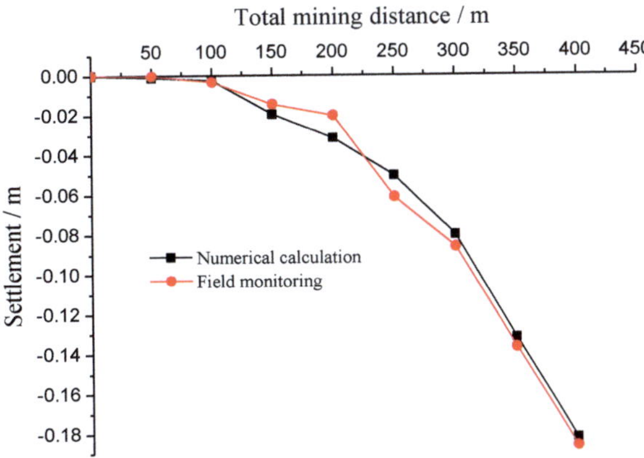

Fig. 9. Comparison between numerical calculation and field monitoring results

6 Conclusion

1. Deformation rules of 1000 kV UHV tower foundation influenced by underground coal mining in adverse rock slope are obtained as well as the potential maximum deformation values.
2. The modified reinforced concrete slab foundation combined with protection panel and length-adjustable anchor bolts can be used to guarantee the safety of the UHV transmission tower in mining areas.
3. The tower foundation deformation is dominated by the effect of coal mining rather than rock slope sliding. The potential moving track of tower foundation surface points is incomplete "C" shape in adverse slope situation.

Acknowledgments. The authors sincerely thank the following agents for their financial supports: National Natural Science Foundation of China (41472259, 51274209); The 13th Five-year Period National Key Research and Development Program of China (2016YFC080250504).

References

Fengli, Y., et al.: Assessment on the stress state and the maintenance schemes of the transmission tower above goaf of coal mine. Eng. Fail. Anal. **31**, 236–247 (2013). https://doi.org/10.1016/j.engfailanal.2013.02.001. Elsevier

Guanglin, Y., et al.: The anti-deformation performance of composite foundation of transmission tower in mining subsidence area. Procedia Earth Planet. Sci. **1**, 571–576 (2009). https://doi.org/10.1016/j.proeps.2009.09.091. Elsevier

Guoqing, H., et al.: Mine Subsidence. China University of Mining and Technology press, Xuzhou (1994). ISBN 9787810214490

Khanal, M., et al.: Prefeasibility study—Geotechnical studies for introducing longwall top coal caving in Indian mines. J. Min. Sci. **50**, 719–732 (2014). https://doi.org/10.1134/S1062739114040139. Springer

Kratzsch, H.: Measures to reduce mining damage. Min. Subsid. Eng. (1983). https://doi.org/10.1007/978-3-642-81923-0_15. Springer

Liang, L., et al.: Evaluation theory and application of foundation stability of new buildings over an old goaf using longwall mining technology. Environ. Earth Sci. **75**, 763 (2016). https://doi.org/10.1007/s12665-016-5574-9. Springer

Likar, J., et al.: Analysis of geomechanical changes in hanging wall caused by longwall multi top caving in coal mining. J. Min. Sci. **48**, 135–145 (2012). https://doi.org/10.1134/S1062739148010157. Springer

Ping, X., et al.: Safety analysis of building foundations over old goaf under additional stress from building load and seismic actions. Int. J. Min. Sci. Technol. **24**, 713–718 (2014). https://doi.org/10.1016/j.ijmst.2014.03.030. Elsevier

Qianjin, S., et al.: Limits to foundation displacement of an extra high voltage transmission tower in a mining subsidence area. Int. J. Min. Sci. Technol. **22**, 13–18 (2012). https://doi.org/10.1016/j.ijmst.2011.07.002. Elsevier

Wenxiu, L. et al.: Ground movements caused by deep underground mining in Guan-Zhuang iron mine, Luzhong, China. Int. J. Appl. Earth Observation Geoinf. **12**, 175–182 (2010). https://doi.org/10.1016/j.jag.2010.02.005. Elsevier

Zhaoping, M., et al.: Deformation, failure and permeability of coal-bearing strata during longwall mining. Eng. Geol. **208**, 69–80 (2016). https://doi.org/10.1016/j.enggeo.2016.04.029. Elsevier

Numerical Analysis of the Seismic Behavior of Alpamarca Tailings Dam in Peru

Jainor Cabrera Huaman and Celso Romanel[✉]

Department of Civil and Environmental Engineering, Pontifical Catholic
University of Rio de Janeiro, Rio de Janeiro, Brazil
romanel@puc-rio.br

Abstract. This paper presents an investigation of the seismic behavior of the
Alpamarca tailings dam which is being raised from elevation 4603 m to 4670 m
above the sea level in a region of high seismic activity in Peru. A seismic hazard
assessment was carried out taking into account the attenuation law recommended
for Peru's geology and the design earthquake was generated through a spectra
matching method. The seismic responses of the dam were computed using FLAC
2D software and the results are presented in terms of accelerations and permanent
displacements. In order to observe the influence of the duration of the earthquake,
the results computed with the total record of accelerations (218 s) were compared
with those determined using a significant duration D_{5-95} = 99.36 s.

1 Introduction

Analysis of structures situated in seismic areas requires an accurate representation of
the effects of earthquakes, a difficult and complicated task due to the random nature of
seismic events. A full investigation involves several studies concerning material
characterization, local seismic hazard assessment, matching acceleration response
spectra, generation of design earthquakes and the execution of dynamic analysis by
numerical methods in the time domain, taking into account the complex nonlinear
response of earth structures excited by seismic loadings.

In this paper the seismic behavior of the Alpamarca tailings dam situated in the
central region of Peru, an area of high seismic activity, is numerically investigated. The
starter dyke is located 4,603 m above the sea level with downstream (1.5H:1.0 V) and
upstream (2H:1 V) slopes. The raising of the tailings dam, built using the downstream
method, is represented through a staged construction involving 10 consecutive layers,
as shown in Fig. 1, and numerically represented in Fig. 2. The geotechnical properties
of the bedrock foundation, the starter dyke and the tailings material are listed in
Table 1, obtained through laboratory (conventional triaxial compression) and field
(seismic refraction) tests.

© Springer International Publishing AG, part of Springer Nature 2019
W. Frikha et al. (eds.), *New Developments in Soil Characterization and Soil Stability*,
Sustainable Civil Infrastructures, https://doi.org/10.1007/978-3-319-95756-2_9

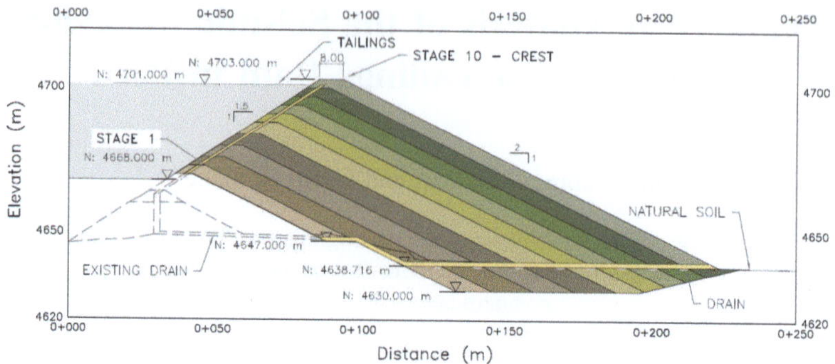

Fig. 1. Staged construction of the tailings dam by the downstream method

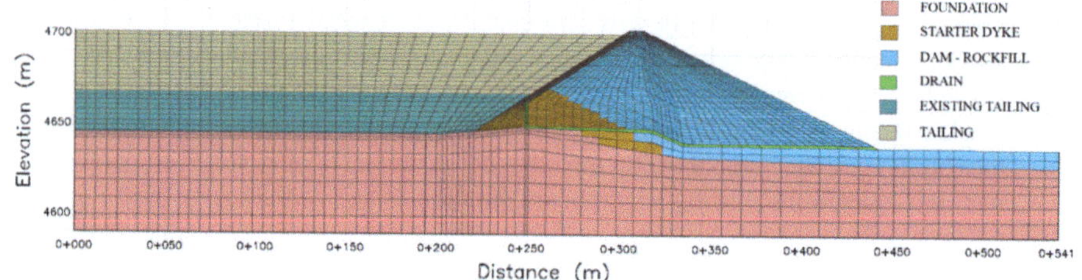

Fig. 2. Numerical model of the dam and tailings deposit

Table 1. Material properties

Material	Unit weigth kN/m3	Compressibility		Shear strength	
		Bulk modulus (MPa)	Shear modulus (MPa)	Cohesion (kPa)	Friction angle(°)
Starter dyke	23.0	880	360	0	38
Rock foundation	25.0	830	380	–	–
Tailings deposit	18.0–20	59–86	19–21	0	28

2 Probabilistic Seismic Hazard Assessment

Peru is located on the Western Coast of the South American crust plate, which is one of the most seismically active regions in the world. The process of subduction of the Nazca plate under the South American plate is the main source of large and destructive earthquakes in Peruvian territory; hence, a proper understanding of the local seismic hazard is necessary for any civil engineering work in the country.

In order to carry out a probabilistic seismic hazard assessment (Cornell 1968) the seismogenic sources nearby the dam site were identified from the earthquake catalog of Peru (Fig. 3). The assessment was carried out using the CRISIS 2015 software (Ordaz et al. 2015), whose results were also compared to those obtained with the web-based OpenQuake platform (GEM 2015).

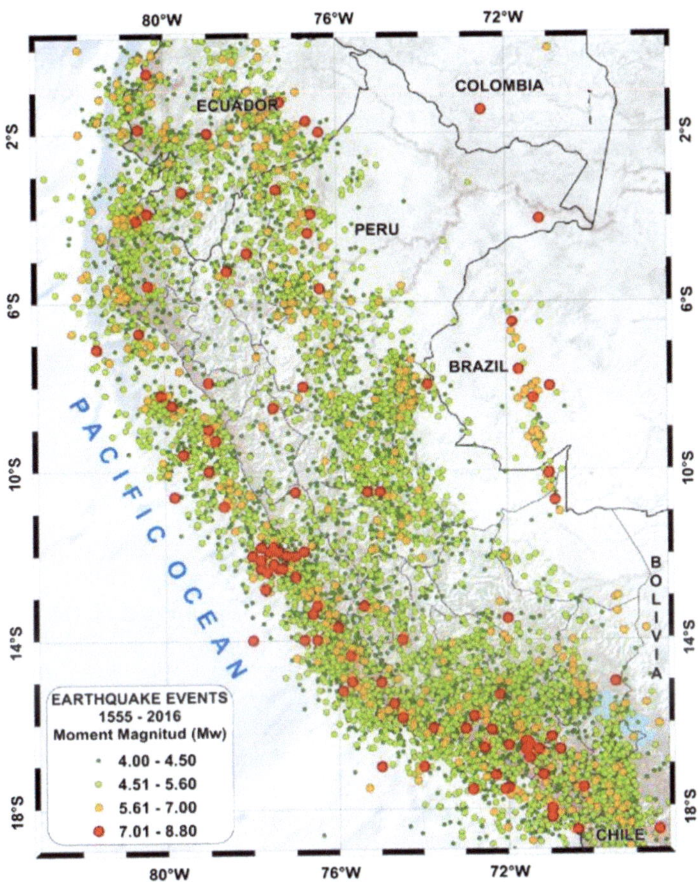

Fig. 3. Earthquake catalog of Peru (Aguilar 2017)

Typically, a probabilistic seismic hazard analysis displays the relative contributions to the hazard assessment due to different values of random components such as the earthquake magnitude and the distance of the source to the site. These results, computed for each seismogenic source separately, are successively combined (disaggregation) for all the seismogenic sources considered in the investigation. The disaggregation of hazard levels into relative contributions from different sources and earthquake events yields a better characterization of the ground motion that may be expected at the site.

The seismic hazard assessment is expressed in terms of annual exceedance probability, and the spectral acceleration is given within a range of peak ground acceleration (PGA) periods of 0–3 s, generally considering a critical damping ratio of 5%.

Figure 4 shows the uniform hazard spectra obtained with both CRISIS 2015 and OpenQuake software. According to the characteristics of tailings dam and the Safety Evaluation Earthquake (SEE) condition recommended by ICOLD (2010) for dam design, the potential collapse of Alpamarca dam is classified as of low consequences, with a return period of 975 years (5% probability of being exceeded in 50 years) being admitted.

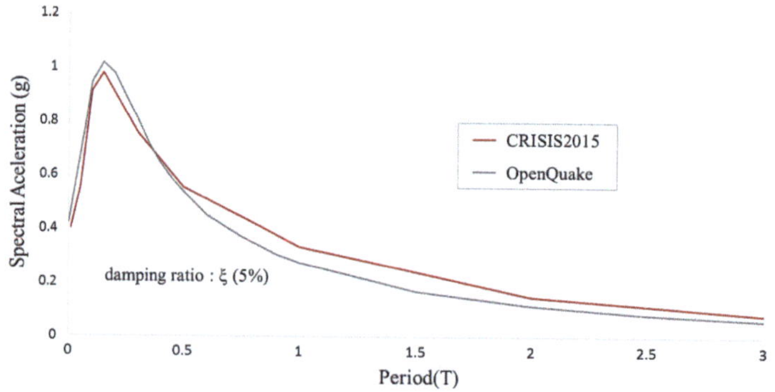

Fig. 4. Uniform hazard spectra considering a return period of 975 years

3 Earthquake Ground Motion

The accelerogram records should incorporate the characteristics of local seismicity. The spectral matching method (Abrahamson 1993; Hancock et al. 2006) was used to generate artificial accelerograms whose response spectra should be consistent with those determined in the seismic hazard assessment.

For the dynamic analysis, a set of three or more earthquake records near the dam site is recommended for the analysis of the structure behavior in order to take into account the variability of time histories and to compensate the peaks and valleys of the different spectra with respect to the target (NIST 2011). These records should be selected considering the duration of the event, tectonic configuration, distance from source to site, frequency content and site conditions. A total of three Peruvian earthquake records were selected (Table 2) and then adjusted in the frequency domain to obtain the best possible match to the spectra design.

Table 2. Selected earthquakes for seismic analysis

Event name	Date	Magnitude	Failure mechanism	Station name
Pisco	August 15, 2007	8.0 Mw	Subduction interface	UNICA (San Luis Gonzaga Univertsity)
Lima	October 03, 1974	8.1 Mw	Subduction interface	Parque de la Reserva
Peru coast	May 31, 1970	8.0 Mw	Subduction intraslab	Parque de la Reserva

4 Static Analysis: Safety Factor

The static behavior of the dam was also evaluated in terms of the safety factor against slope failure, using the method of shear strength reduction. The advantage of using a numerical method for calculation of the safety factor is that the shape and location of the potential failure surfaces need not be previously defined, as in the limit equilibrium methods but, on the other hand, caution is required in the interpretation of the results because often the collapse detected by the computational program refers to local landslides (near the crest of the dam or superficial failures along the slopes) that do not compromise the overall stability of the structure. Figure 5 shows the potential failure surface after the placement of the last layer (10th phase of the staged construction) corresponding to a static safety factor FS = 1.57.

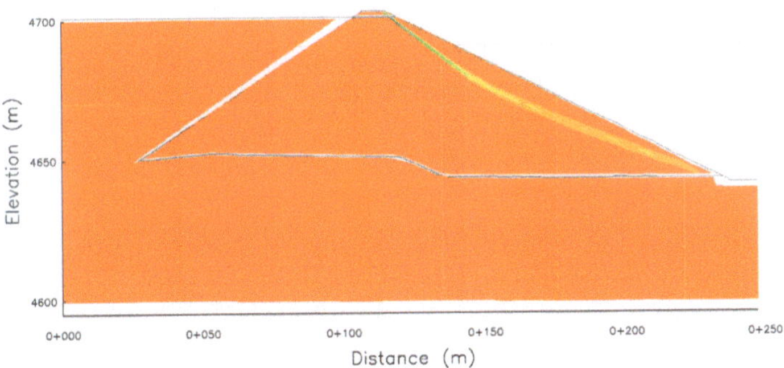

Fig. 5. Potential failure surface with static safety factor FS = 1.57

5 Dynamic Analysis

5.1 Shear Modulus and Hysteretic Damping

The bedrock was represented as a linear elastic material with shear modulus determined from S-wave velocity measured in geophysical tests. For the materials of the body of the dam the maximum shear modulus G_{max} was calculated as a function of the effective octahedral normal stress, according to Eq. 1 proposed by Seed and Idriss (1970), Seed et al. (1984):

$$G_{max} = 21.7 \left(K_{2,max} \right) p_a \left(\frac{\sigma'_m}{p_a} \right)^{0.5} \tag{1}$$

where σ'_m is the effective octahedral stress, p_a the atmospheric pressure and the value of parameter $K_{2,max}$ varies between 30 and 70, obtained from specific tables (Seed and Idriss 1970) as a function of the void ratio or relative density.

The equivalent-linear model simulates non-linear cyclic behavior including shear modulus degradation with shear strain and strain-dependent damping ratio (hysteretic damping). The inclination of each loop is dependent on the secant shear modulus (G) and the breadth of the hysteresis loop is related to the area inside the loop, a measure of energy dissipation. The slope of the backbone curve at the origin corresponds to the maximum tangent shear modulus (G_{max}) but at greater cyclic shear strain amplitudes the modulus ratio G/G_{max} will drop to values less than one. This variation of shear modulus ratio with shear strain is represented by a modulus reduction curve, used along with the shear yield strength given by the Mohr-Coulomb failure criterion.

The modulus degradation curves determined from laboratory tests by Seed et al. (1984) for rockfill, and by Winckler (2014), for fine tailings material, were used in this research. These degradation curves were fitted to a 3-parameter sigmoidal function available in FLAC 2D (Itasca Consulting Group 2016).

The breadth of the hysteresis loops also increases with increasing cyclic shear strain amplitude. In FLAC 2D code, the damping ratio ξ is obtained on the basis of the theoretical Eq. 2:

$$\xi = \frac{\Delta W_d}{4\pi W_s} \tag{2}$$

where ΔW_d is energy dissipated in one cycle of loading and W_s the maximum strain energy stored during the cycle.

As a consequence, the modulus degradation curve agrees with that available in other geotechnical software (such as SHAKE 2000) but the curves of the damping ratio increase are not generally coincident, since in FLAC 2D the damping ratio is theoretically calculated (Eq. 2) while in most other computer programs the damping ratio curves, obtained from laboratory tests, are directly specified.

Due to the inconsistency observed in the damping ratio curves in the entire range of cyclic shear strain, it is suggested to match the theoretical and laboratory curves within a specific strain range of interest. In this research, the fit process considered the strain interval between 10^{-5} to 10^{-3}.

5.2 Spectral Matching

Three seismic records were pre-processed by a baseline correction and the application of a Butterworth bandpass filter in the range of 0.1 to 15 Hz. Subsequently, spectral matchings were carried out with the SeismoMatch software (2016), based on the algorithm proposed by Abrahamson (1993) and Hancock et al. (2006), adding wavelets to reduce the difference between the original and the adjusted spectra. Figure 6 shows the initial spectrum from the earthquake acceleration records, the corresponding matched spectrum and the uniformly probable hazard spectrum (return period of 475 years). Figure 7 illustrates the artificially generated design earthquakes with peak horizontal acceleration $PHA^{rock} = 0.4$ g in the bedrock.

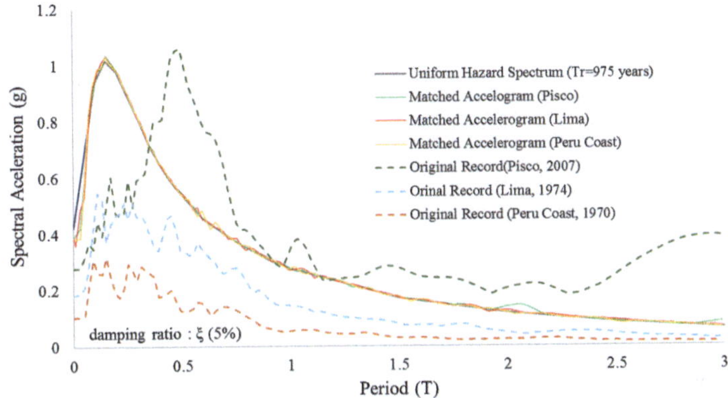

Fig. 6. Target, original and matched acceleration response spectra

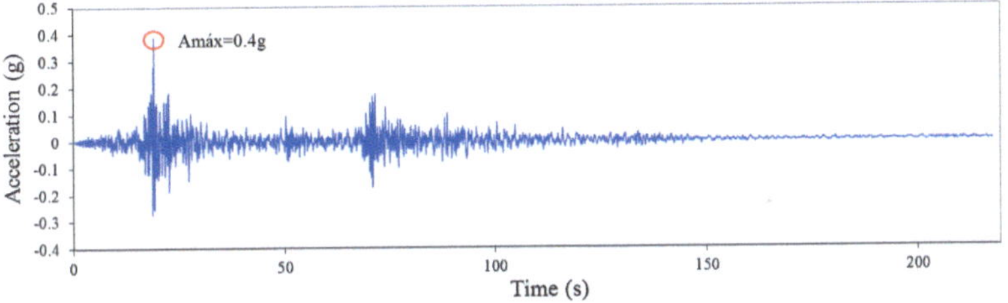

Fig. 7. Matched acceleration time-history

5.3 Seismic Response

Figure 8 shows the seismic responses at points in the tailings deposit and in the dam body. They are compared to the spectrum of the design earthquake (input motion applied to the base of the numerical model). The corresponding time acceleration records are shown in Fig. 9. The respective acceleration power spectra were also constructed through the Fast Fourier Transform (FFT), which allowed the identification of the predominant frequencies in the response computed at each point (Fig. 10).

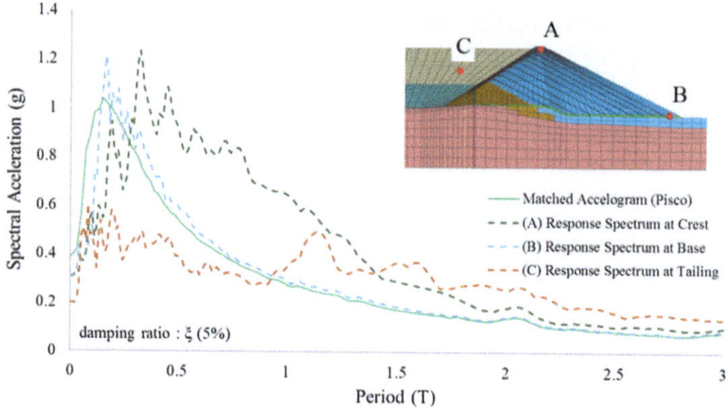

Fig. 8. Comparison between acceleration response spectra at points A, B, C

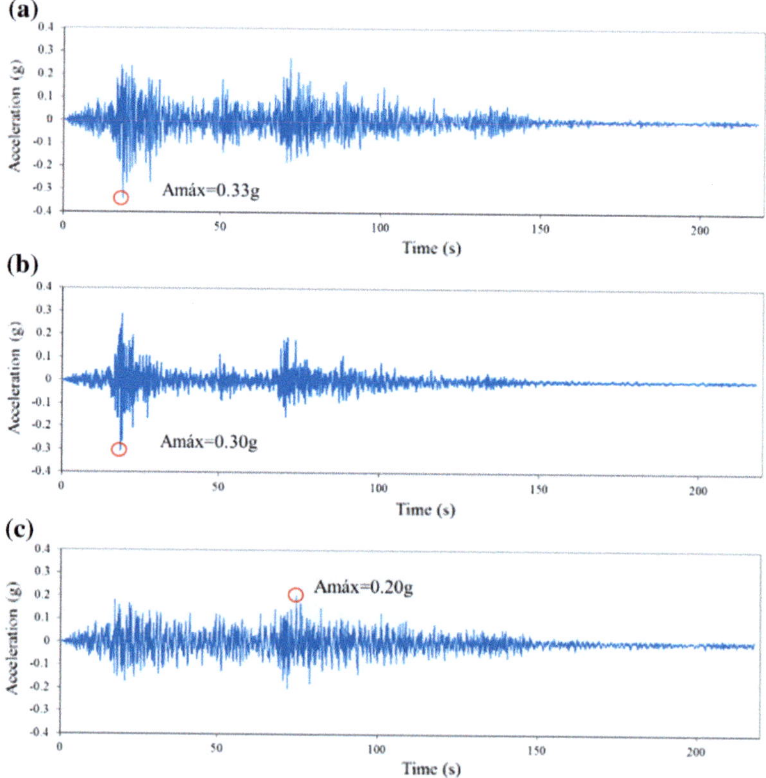

Fig. 9. Horizontal accelerations at points A, B, C

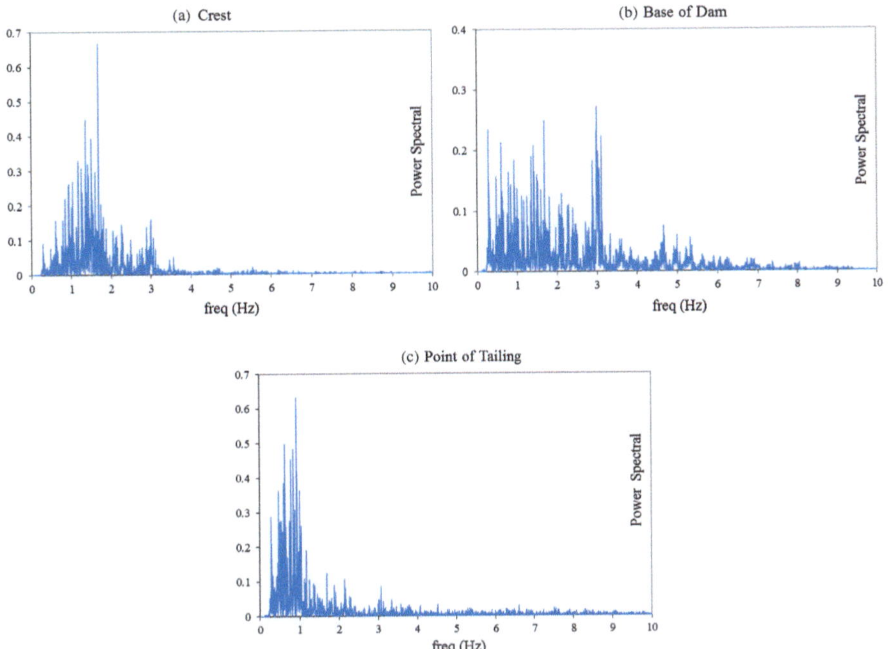

Fig. 10. Power spectrum of accelerations at points A, B, C

5.4 Permanent Displacements

The permanent horizontal displacements predicted in the downstream slope reached values up to 0.50 m, while the maximum displacements in the tailings deposit were observed up to 3.75 m (Fig. 11). With respect to the distribution of permanent vertical displacements (Fig. 12), the maximum predicted value was 1.20 m.

5.5 Duration of Seismic Record

An important factor in seismic analysis is the choice of the time interval of accelerations because usually only the highest intensity range is considered in order to decrease the processing time. A large number of definitions in the literature for the strong ground motion durations are based upon the energy content of the accelerograms. The significant duration is usually based on the integral of the square of the acceleration time history, although Kempton and Stewart (2006) have also used the square of the velocity time history and Trifunac and Brady (1975) the squares of both the velocity and the displacement time history.

The significant duration is based on the Arias intensity I_A (1969), defined as the integral of the square of the acceleration times a constant ($\pi/2\ g$), as shown in Eq. 3.

$$I_A = \frac{\pi}{2g} \int_0^T a^2(t)\,dt \tag{3}$$

where a(t) is the acceleration time history and T is total duration of the acceleration time history.

A plot h(t) that portrays the buildup of this energy with time for a strong motion record is known as a Husid plot (1969), in which the Arias intensity is normalized between the values 0 and 1 according to Eq. 4.

$$h(t) = \frac{\int_0^T a^2(t)\,dt}{\int_0^t a^2(t)\,dt} \tag{4}$$

Trifunac and Brady (1975) suggested that the interval between 5% and 95% of the normalized Arias intensity is an appropriate choice for the significant duration, although in the literature the time interval 5–75% is also a commonly used significant duration.

The Pisco earthquake presented a total duration of 218 s and a significant duration $D_{5-95} = 99, 36$ s between $t_5 = 3, 27$ s e $t_{95} = 102, 63$ s. The permanent displacements computed with the signification duration of Pisco earthquake have shown satisfactory agreement with the previously computed results considering the full acceleration time history, as indicated in Table 3. The distributions of permanent horizontal and vertical displacements for both cases may be also compared in Figs. 11, 12, 13 and 14.

Table 3. Permanent displacements determined with full acceleration time history (218 s) and significant duration (D_{5-95})

Earthquake duration	Horizontal displacement (m)		Vertical displacement (m)	
	Tailing	Slope	Tailing	Slope
218 s	−2.7	0.60	−0.8	−0.20
99.36 s	−1.80	0.60	−0.5	−0.20

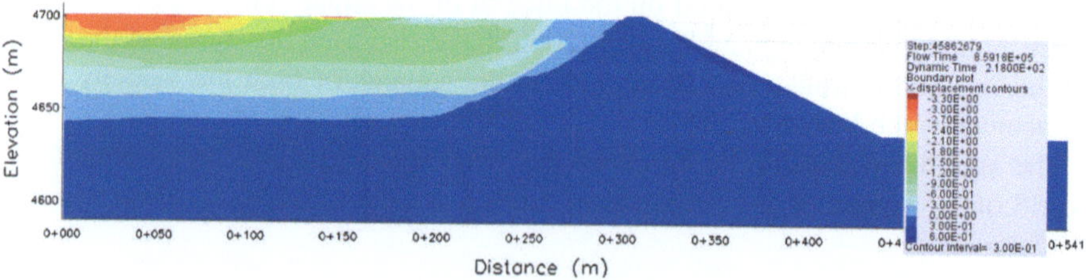

Fig. 11. Distribution of permanent horizontal displacements (218 s)

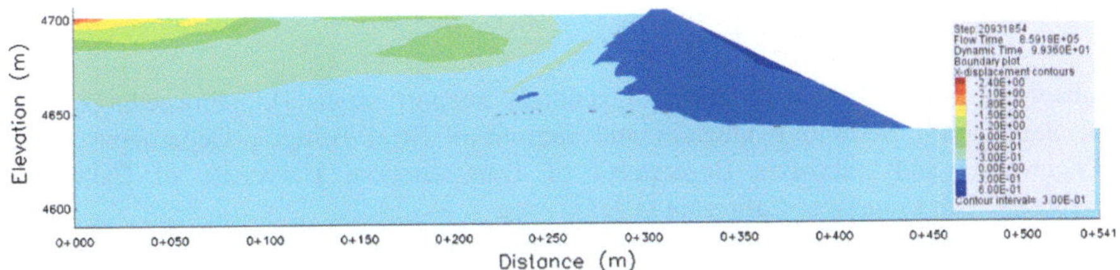

Fig. 12. Distribution of permanent horizontal displacements ($D_{5-95} = 99.36$ s)

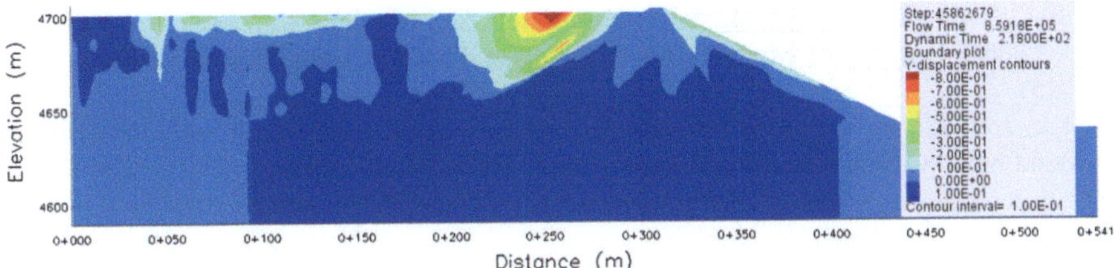

Fig. 13. Distribution of permanent vertical displacements (218 s)

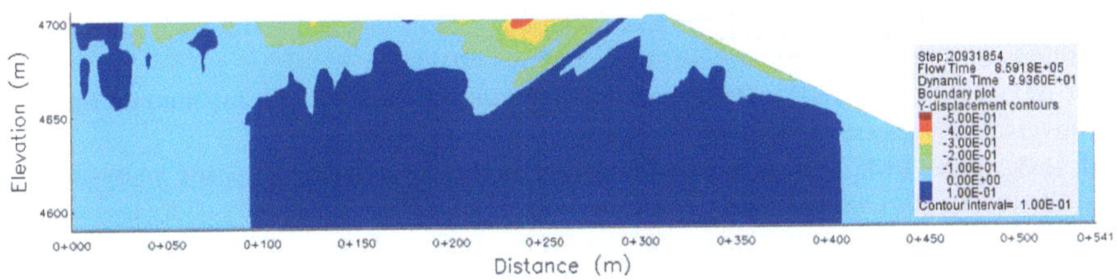

Fig. 14. Distribution of permanent vertical displacements ($D_{5-95} = 99.36$ s)

6 Conclusion

The main objectives of this paper were the seismic hazard assessment and the analysis of the dynamic behavior of the Alpamarca tailings dam, located in a region of high seismic activity in Peru. The seismic hazard assessment is fundamental to establish with better confidence the design earthquakes, carried out in this research by means of the spectral matching method. One of the difficulties in estimating the seismic hazard in a region where data from seismic instrumentation is deficient is the lack of a proper attenuation laws for seismic movements. In Peru, the current state of practice it the use of attenuation laws developed for North America, considering the similitude of the existing geological conditions.

One of the points to be emphasized is the need to consider an adequate evaluation of the duration for a strong motion record which may lead to a significant reduction in computer processing time. In this investigation, permanent displacements obtained with the Pisco full acceleration time history (218 s) and the significant duration ($D_{5-95} = 99.36$ s) presented good agreement in distribution and maximum values.

References

Abrahamson, N.A.: Spatial variation of multiple support inputs. In: Proceedings, 1st U. S. Seminar on Seismic Evaluation and Retrofit of Steel Bridges, Department of Civil Engineering and California Department of Transportation, University of California at Berkeley, San Francisco, California (1993)

Aguilar: Probabilistic seismic hazard assessment in the peruvian territory. In: Proceedings of the 16th World Conference on Earthquake Engineering, Santiago–Chile (2017)

Arias, A.: A measure of earthquake intensity. In: Hansen, R.J. (ed.) Seismic Design for Nuclear Power Plants, pp 438–483. MIT Press (1969)

Cornell, C.A.: Engineering seismic risk analysis. Bull. Seismol. Soc. Am. **58**, 1583–1606 (1968)

GEM—Global Earthquake Model: The OpenQuake Platform https://platform.openquake.org/ (2015)

Hancock, J. et al.: An improved method of matching response spectra of recorded earthquake ground motion using wavelets. J. Earthq. Eng. **10**(1): 67–89 (2006)

Husid, L.R.: Características de Terremotos, Análisis General. Revista del IDIEM 8, Santiago, Chile, pp 21–42 (1969)

ICOLD: Selecting seismic parameters for large dams. Bulletin 72 (2010)

Itasca Consulting Group: FLAC—Fast Lagrangian Analysis of Continua, vol. 8 (2016)

Kempton, J.J., Stewart, J.P.: Prediction equations for significant duration of earthquake ground motions considering site and near-source effects. Earthq. Spectra **22**(4), 985–1013 (2006)

NIST—National Institute of Standards and Technology: Selecting and scaling earthquake ground motions for performing response-history analyses (2011)

Ordaz, M., et al.: CRISIS2015: Program for Computing Seismic Hazard. Instituto de Ingeniería, Universidad Nacional Autónoma de México, Mexico (2015)

Seed, H.B., Idriss, I.M.: Soil moduli and damping factors for dynamic response analysis, Report n. EERC 70–10, University of California, Berkeley (1970)

Seed, H.B. et al.: Moduli and damping factors for dynamic analysis of cohesionless soils, Report n. EERC 84–14, University of California, Berkeley (1984)

Seismosoft Ltd: SeismoMatch's Help System (2016)

Trifunac, M.D., Brady, A.G.: A Study on Duration of Strong Earthquake Ground Motion. Bull. Seismol. Soc. Am. **65**, 581–626 (1975)

Winckler, C. et al.: CPTu-Based State Characterization of Tailings Liquefaction Suscetibility, Proceedings of the 34th United States Association of Dams (USSD), San Francisco (2014)

Identifying the Defects Presented on the Exterior Layers of a Structure by Employing 3D Point Clouds and Thermography

Chi-Ping Wang[1], Yishuo Huang[2]([⊠]), Sung-Chi Hsu[2], and Jia-Jian Hong[2]

[1] Department of Civil Engineering, Feng Chia University, Taichung, Taiwan
[2] Department of Construction Engineering, Chaoyang University of Technology, Taichung, Taiwan
yishuo@cyut.edu.tw

Abstract. Severe weathers and climate changes do have profound impacts on the exterior layer of structures. Especially for those defects presented on the exterior layers of buildings, those defects can make the exterior layers of buildings worse, and, sometimes, the life cycles of buildings are decreased. How to identify those defects shown on the exterior layers of structures is an important issue for the sustainable structure management. Non-destructive testing (NDT) methods have been widely applied to detect those defects presented on the A terrestrial layers. However, NDT methods cannot provide efficient and reliable information for identifying the defects because of the huge examination areas. Thermography is a product of a thermal infrared camera. Thermography is used to record the surface temperature information, and the differences of the recorded surface temperatures are small such that it was difficult to analyze the collected thermographs. The defects presented on the exterior layer of buildings are usually located at the places whose surface temperatures are higher than their corresponding neighbors. This paper proposed to employ image segmentation to cluster those pixels with similar surface temperatures such that the thermography can be composed of the limited groups. For each segmented group, the surface temperature distribution in each segmented group is homogenous. In doing so, the regional boundaries of the segmented regions can be identified and extracted. A terrestrial laser scanner (TLS) is widely used to collect the point clouds of a structure, and those point clouds can be modeled as a 3D structure. This study established a mapping model such that the segmented thermography can be projected on the 3D structure. In doing so, the 3D structure provides the defects and their spatial locations. In this paper, the structure (a wind turbine) located at Taichung County Gaomei Wetlands is used as an example.

© Springer International Publishing AG, part of Springer Nature 2019
W. Frikha et al. (eds.), *New Developments in Soil Characterization and Soil Stability*, Sustainable Civil Infrastructures, https://doi.org/10.1007/978-3-319-95756-2_10

1 Introduction

Structure health monitoring (SHM) is defined as a process to detect and characterize the damages occurring on an artificial structure, and it becomes an important issue to sustainable structure managements. Traditionally, the damages occurring on a structure can be identified by employing destructive and nondestructive testing methods; usually, nondestructive testing methods are preferred because they can identify the damages without causing extra damages on a structure. There are many nondestructive testing (NDT) methods have widely been employed to evaluate the aged buildings because of their reliability and effectiveness. NDT methods can not only permanently alter the inspected objects but also provide the defect information. In general, NDT methods can be classified into Visual Inspection, Proof Load Test, Vibration Testing, Impact Testing, Ultrasonic NDT, Conductivity and Radar (McCann and Forde 2001). The visual inspection mentioned above is the most economical way to detect the damages. However, for a large structure (like wind turbines), visual inspection is hard to apply to locate the damages. Recently, Unmanned Aerial Vehicle (UAV) is employed to examine the structures instead of applying the visual inspection because the examined structures are usually huge. Phung et al. employed the camera installed on UAV to develop an optimal flying path to examine the surface conditions of buildings and bridges (Phung et al. 2017); Moss et al. demonstrated that structural health monitoring systems could be based on a UAV system (Moss et al. 2017). Terrestrial laser scanning (TLS) is an innovate technology to collect the spatial information of a structure, and it is based on the transmission and receiving of pulsed light (Chen 2012). Yang et al. employed TLS to collect the deformation information of concrete composite structures (Yang et al. 2017); Liu applied the TLS to monitor the deformation behaviors of bridges (Liu 2010). Both systems offer huge spatial data such that the structure health conditions can be monitored.

Infrared thermography is the product of remote sensing techniques and is usually used to record the surface temperature of the observed objects. Infrared thermography is widely used to identify defects presented in objects by observing the radiant heat pattern (Huang and Wu 2010). For a thermal camera, radiation is converted to temperature depending on the emissivity value of the specimen. Emissivity values are in the range of 0 and 1, and they are defined as the ratio of a body's emission spectrum at a given temperature to that of a blackbody at the same temperature (Plotnikov and Winfree 1998). If the emissivity value is 1, it means that the test body is a black body; if it is 0, the test body completely reflects all the received energy. Infrared thermal cameras are designed to record the temperatures transformed from emissivity values. Changes in the recorded temperatures on thermal images reveal possible surface flaws (Maldague 2001). Furthermore, those areas in which the surface temperature information is different with their surrounding neighborhoods can be used to indicate the locations of the potential defects presented on objects.

The paper integrates the spatial information collected by employing both UAV and TLS and surface temperature information recorded by a thermal infrared camera to identify the defect information of structures. The spatial information of a structure is too abundant to handle. Hence, a 3D model is built to simplify the spatial information collected from UAV and TLS. Then, the thermal images are analyzed by segmenting each thermal image into a series of sub-regions such that the surface temperature distribution in each segmented region is homogeneous. In doing so, the surface temperature information in each segmented region is illustrated such that the temperature differences can be illustrated and used to identify the defect locations. Eventually, the segmented regions are treated as texture information and attached on the 3D model such that the textured model provides defect information.

A wind turbine located at Taichung County Gaomei Wetlands is used as an example. The remainder of this paper is organized as follows. In the next section, the ways to build the 3D model of the wind turbine is introduced, and multilayer level set model used to analyze the thermal images are briefly introduced. In Sect. 3 illustrates the 3D model and the segmented results by employing the multilayer level set approach on real thermal infrared images. Eventually, some conclusions and related discussions are provided.

2 Hyper Approach by Integrating Different Point Clouds and Thermography

Structural health monitoring systems usually adopt a system to monitor a structure. Both UAV and TLS used in SHM can offer spatial information of the monitored structures. However, UAV provides better spatial information illustrated on the top of the monitored structures; oppositely, TLS offers better spatial information shown in the structure except for the top because the top information is blocked. Hence, the paper employed the point clouds from UAV and TLS to build the 3D model of the structure. The way to generate the 3D model is introduced. Then, thermal infrared cameras were employed to record the surface temperature information. The collected images were analyzed by applying the multilayer segmentation such the recorded surface temperature information can be grouped into few groups such that the surface temperature information in each group can be replaced by the average temperature of the grouped region. The multilayer segmented approach is briefly introduced.

2.1 3D Model Generated from Different Point Clouds

Point clouds are the products by using UAV or TLS to collect the spatial data of the monitored structure. Both of them will have their coordinate systems such that the collected data from two systems cannot be directly combined. In this paper, ground control points were established such that the collected data by UAV or TLS could be projected to the same coordinate system. The projection between the collected and controlled points is assumed that the projection can be explained by a polynomial function. The corresponding points of the ground control points are chosen from the point clouds; the optimal locations are determined by weighting the possible points and their neighborhood from the given point clouds. In doing so, two point clouds from UAV and TLS are combined. The combined point cloud is imported and transformed into the vector format to build the 3D model of the monitored structure.

2.2 Analyzing Thermography by Applying the Multilayer Segmentation

The fundamental idea of applying image segmentation on thermography is to partition a given thermal infrared image into a series of sub-regions such that each sub-region is homogeneous. In doing so, the surface temperature distributions can be identified and located in the thermal infrared image. Multilayer segmentation proposed by Chung and Vese segments the given thermal infrared image according to the pre-defined thresholds such that the pixel values shown in the segmented regions are grouped and centered by the pre-selected thresholds (Chung and Vese 2010). In the multilayer approach, let I_0 be a thermal image and the two level set functions are used to segment the image into a series of sub-regions with distinct level values $\{l_1 < l_2 < \cdots < l_m\}$ and $\{k_1 < k_2 < \cdots < k_n\}$ for ϕ_1 and ϕ_2 the level set functions, respectively. Then the energy functions generated by level set functions for ϕ_1 and ϕ_2, pre-selected thresholds $\{l_1 < l_2 < \cdots < l_m\}$ and $\{k_1 < k_2 < \cdots < k_n\}$, and the regional constants c (the average values of the segmented regions) can be defined as follows (Chung and Vese 2010):

$$
\begin{aligned}
(c, \varphi) = &\sum_{i,j=1}^{m-1,n-1} \int_\Omega |I_0 - c_{i,j}|^2 \chi_1 dx + \sum_{i=1}^{m-1} \int_\Omega |I_0 - c_{i,0}|^2 \chi_2 dx + \sum_{i=1}^{m-1} \int_\Omega |I_0 - c_{i,n}|^2 \chi_3 dx + \sum_{j=1}^{n-1} \int_\Omega |I_0 - c_{0,j}|^2 \chi_4 dx \\
&+ \sum_{j=1}^{n-1} \int_\Omega |I_0 - c_{m,j}|^2 \chi_5 dx + \int_\Omega |I_0 - c_{0,0}|^2 \chi_6 dx + \int_\Omega |I_0 - c_{0,n}|^2 \chi_7 dx + \int_\Omega |I_0 - c_{m,0}|^2 \chi_8 dx + \int_\Omega |I_0 - c_{m,n}|^2 \chi_9 dx \\
&+ \mu \sum_{i=1}^{m} \int_\Omega |\nabla H(\varphi_1 - l_i)| dx + \mu \sum_{j=1}^{n} \int_\Omega |\nabla H(\varphi_2(x) - k_j)| dx
\end{aligned}
$$

$$(1)$$

where H is the Heaviside function, $\mu > 0$ is a weight parameter, and c_{ij} is the sub-regional constant. The parameter χ_i is the combinations of level set functions ϕ_1 and ϕ_2; for an example, χ_1 is defined as $H(\phi_1 - l_i)H(l_{i+1} - \phi_1)H(\phi_2 - k_j)H(k_{j+1} - \phi_2)$. As for other parameters χ_i, those parameters can be referred the paper of Huang et al. (2014). The optimal approximation I derived from the segmented results can be shown as:

$$
\begin{aligned}
I = &\sum_{i=1}^{m-1} \sum_{j=1}^{n-1} c_{ij}\chi_1 + \sum_{i=1}^{m-1} c_{i0}\chi_2 + \sum_{i=1}^{m-1} c_{in}\chi_3 + \sum_{j=1}^{n-1} c_{0j}\chi_4 \\
&+ \sum_{j=1}^{n-1} c_{mj}\chi_5 + c_{00}\chi_6 + c_{0n}\chi_7 + c_{m0}\chi_8 + c_{mn}\chi_9
\end{aligned}
$$

$$(2)$$

The level set functions ϕ_1 and ϕ_2 can be found by employing finite difference such that initial level set functions will change their shapes to meet the energy minimization steps by steps by using iteration scheme (Chung and Vese 2010; Huang and Wu 2010; Huang et al. 2014).

3 Processed Results by Employing the Proposed Approach

A wind turbine located at Taichung County Gaomei Wetlands was chosen because the structure is hard to approach. The spatial data collected by UAV is illustrated in Fig. 1. Similarly, TLS is used to collect the spatial data, and the results are illustrated in Fig. 2. The 3D model of the wind turbine is generated by employing the point clouds collected by applying UAV and TLS, and the 3D model is presented in Fig. 3. As for thermography, NEC Thermo Gear-G120 and FLIR VUE were used to record the surface temperature information of the wind turbine. Thermo Gear-G 120 was installed on the ground to measure the surface temperature information, and FLIR VUE was installed on UAV to collect a series of images. With the multilayer segmentation, the thermal images were segmented into few regions. The processed results are illustrated in Figs. 4. and 5, respectively.

(a) Using UAV to Collect Data   **(b) The Point Cloud Generated from UAV**

Fig. 1. The collected spatial data of the wind turbine located at Taichung county Gaomei Wetlands

Fig. 2. The point clouds of the wind turbine collected by employing TLS.

Fig. 3. The generated 3D model of the wind turbine

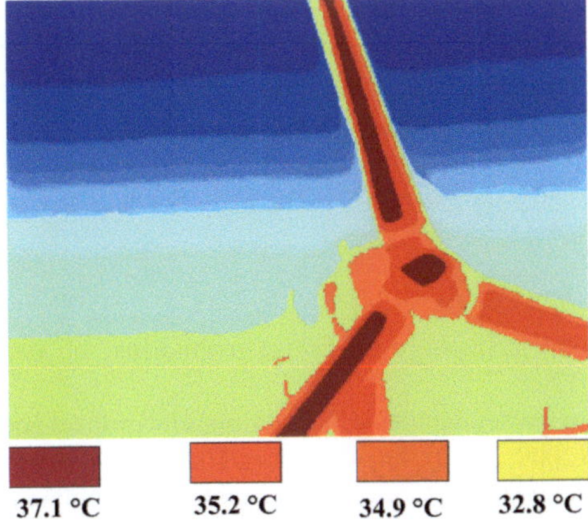

Fig. 4. The segmented results of applying the multilayer segmentation on the thermal image collected by FLIR.

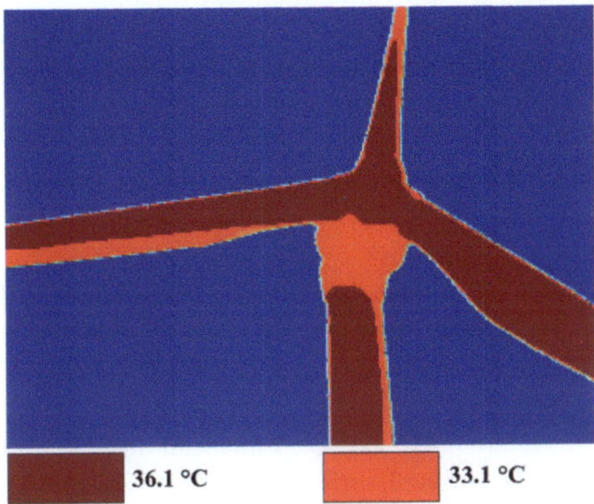

Fig. 5. The segmented results of applying the multilayer segmentation on the thermal image collected by Thermo Gear-G120.

4   Conclusions

The proposed approach employs the 3D model generated by using point clouds from UAV and TLS to illustrate the spatial information of the wind turbine. Then, thermography is used to record the surface temperature information, and the multilayer segmentation is employed to analyze the given thermal images. The thermal camera FLIR VUE cannot directly collect the surface temperatures; hence, the surface temperature information is obtained by comparing the surface temperature information collected by Thermo G-120. The conclusions are summarized as follows:

1. 3D model provides the spatial information of the monitored structure;
2. Thermography offers the surface temperature information of the monitored structure, and the differences of the analyzed surface temperatures are the important clues to identify the defects shown in the structure.

References

Chen, S.E.: Laser scanning technology for bridge monitoring. In: Laser Scanner Technology. Intechopen Com. (2012)

Chung, G., Vese, L.A. Image segmentation using a multilayer level-set approach. Comput. Vis. Sci. **12**(6), 267–285 (2009, 2010). https://doi.org/10.1007/s00791-008-0113-1

Huang, Y., Wu, J.: Infrared thermal image segmentations employing the multilayer level set method for non-destructive evaluation of layered structures. NDT&E Int. **43**(1), 34–44 (2010). https://doi.org/10.1016/j.ndteint.2009.08.001

Huang, Y., Chiu, C.T., Lee, M.G., Lin, S.Y.: Applying the multilayer level set approach to explore the thermal surface characteristics of hot mix asphalt. Constr. Build. Mater. **53**, 621–634 (2014). https://doi.org/10.1016/j.conbuildmat.2013.10.097

Liu, W.: Terrestrial lidar-based bridge evaluation. Ph.D. Dissertation, Department of Civil and Environmental Engineering, The University of North Carolina at Charlotte (2010)

Maldague, X.P.V.: Theory and practice of infrared technology for non-destructive testing. Wiley, New York (2001)

McCann, D.M., Forde, M.C.: Review of NDT methods in assessment of concrete and masonry structure. NDT&E Int. **34**(2001), 71–84 (2001). https://doi.org/10.1016/s0963-8695(00)00032-3

Moss, S.D., Davis, C.E., van der Velden, S., Jung, G., Smithard, J., Rosalie, C., Norman, P., Knowles, M.L., Galea, S.C., Dang, P., Najic, N.: A UAS testbed for the flight demonstration of structural health monitoring systems. Procedia Eng. **188**, 456–462 (2017). https://doi.org/10.1016/j.proeng.2017.04.508

Phung, M.D., Quach, C.H., Dinh, T.H., Ha, Q.: Enhanced discrete particle swarm optimization path planning for UAV vision-based surface inspection. Autom. Constr. **81**, 25–33 (2017). https://doi.org/10.1016/j.autcon.2017.04.013

Plotnikov, Y.A., Winfree, W.P. Advanced image processing for defect visualization in infrared thermography. In: Proceedings of SPIE, the International Society for Optical Engineering, vol. 3361, pp. 331–338. Bellingham (March 1998)

Yang, H., Omidalizarandi, M., Xu, X., Neumann, I.: Terrestrial laser scanning technology for deformation monitoring and surface modeling of arch structures. Composit. Struct. 173–179 (2017). https://doi.org/10.1016/j.compstruct.2016.10.095

Examination of the Ignition Oven Method Correction Factor for Hot Mix Asphalt with Granite Aggregates

Can Chen[1]([✉]), Wolfgang O. Eisenhut[1], and Thanh Ngo[2]

[1] ChemCo Systems Inc., Redwood City, CA, USA
{chen, eisenhut}@chemcosystems.com
[2] Graniterock, South San Francisco, CA, USA
tngo@graniterock.com

Abstract. Ignition oven is a widely-used equipment for determining the asphalt content and gradation in hot mix asphalt. Accurate determination of the asphalt content and post-plant gradation by ignition oven method are the fundamentals of quality control in hot mix asphalt production. However, a correction factor is usually needed, mainly due to the mass loss of the asphalt binder and the aggregates decomposition during the high temperature burn-off process. This study investigated the influences that could affect the correction factors for both asphalt content and aggregate gradation using granite materials. Results show that smaller size aggregates would give a higher asphalt content correction factor than the large size aggregates, and a higher ignition oven temperature would lead to higher asphalt content correction factor. As for the aggregate gradation correction factor, larger size aggregate would lead to more breakdown than the smaller size aggregate, however, for the same type of aggregate, different ignition oven setting temperature does not obviously change the gradation correction factor. The polarized light microscope analysis demonstrates that granite aggregates that subjected to the extremely high temperature in the ignition oven, have minerals being partially decomposed, producing brown fractured chlorite and hornblende altered by chlorite.

1 Introduction

One of the primary quality control tests performed on hot mix asphalt (HMA) is the determination of asphalt binder content and post-plant gradation. Many state highway agencies pay paving contractors based on the accuracy of binder content in the plant-produced mix comparing to the Job Mix Formula (JMF). Some even verify the post-production HMA aggregate gradations against submitted the JMF. Therefore, it is very important for both the contractors and highway transportation agencies that the binder content and gradation determination be accurate as well as timely. The ignition oven method is a fast and simple procedure for asphalt binder content determination without requiring the use of environmentally hazardous chemicals or radioactive sources. The ignition oven heats the asphalt mix to a temperature above the flame point of binder and literally ignites the asphalt binder, leaving behind "clean" aggregate. The loss of weight during this process is monitored, and asphalt binder content is calculated by subtracting

© Springer International Publishing AG, part of Springer Nature 2019
W. Frikha et al. (eds.), *New Developments in Soil Characterization and Soil Stability*,
Sustainable Civil Infrastructures, https://doi.org/10.1007/978-3-319-95756-2_11

the mass of the aggregate remaining after the asphalt binder is burned off from the initial mass of the test sample. After asphalt binder being burned off, "clean" aggregates remain, which can then be washed, dried and tested for gradation accuracy. When using ignition oven to determine the asphalt content of HMA, correction factors should be established in advance, which include both asphalt content correction factor and aggregate gradation correction factor. However, some concern has been expressed regarding the effect of the extremely high temperature of ignition oven test on the correction factor. This study is trying to figure out the influence factors for both asphalt binder and aggregates correction factor based on the AASHTO T 308 method.

Before testing is performed on HMA production asphalt mixture, a mixture correction factor must be determined. The definitions are shown below (AASHTO 2014):

Asphalt content correction factor—Asphalt binder content results may be affected by the type of virgin aggregate and RAP material in the mixture and the ignition furnace. The asphalt binder correction factor is calculated from the difference between the actual and measured asphalt binder contents for each sample.

Aggregate gradation correction factor—Due to potential virgin and RAP aggregate breakdown during the ignition process, an aggregate gradation correction factor will be determined. Aggregate gradation correction factor is determined to by comparing the difference of gradation results on residual aggregate to the prior-burning aggregate gradation values.

The asphalt content correction factor is usually considered to be due to a release of moisture that is trapped in the aggregate and fines. When HMA mixture heated in high temperature, aggregate particles decrepitate due to the presence of water/moisture contained in the minerals. Water inclusions are important in regards to decrepitation. In decrepitation, the water pressure is released which can further result in mineral decomposition that leads to a crack or crystal break-up during a coupled-thermal-physical process. The process has been widely studied by mineralogists on granite and other types of rocks. Using the thermogravimetric analysis (TGA) and the Pillkington method, they noticed that the decrepitation is affected by particle size, sample mass and type of rocks (Smith 1957; McCauley and Johnson 1991; Dollimore et al. 1994). Smith (1957) investigated the decrepitation during the heating of granite aggregate. He noticed that the characteristic shape of decrepitation curve of granite, and presumably of igneous minerals also, is a fairly rapid acceleration soon after the stage begins, followed by a more or less linear increase of rate to a peal 100–200 °C above the beginning, and finally a slower decrease of rate, with continuous deceleration, to many hundreds of degrees above the beginning. However, none of the studies were considered from the engineering aspect and the effect of asphalt binder is not involved in research. In this study, the changes of asphalt binder content and aggregate gradation under high temperature during ignition oven testing was investigated.

2 Objectives

Determine if there is any relationship between ignition oven correction factor (both asphalt content and gradation) to aggregate size, oven setting temperature and asphalt binder content.

3   Materials

Aggregate materials used in this study were collected in the same day sampled from local quarry stockpile to avoid material properties variation. For comparison, this study used two different size aggregates: the coarser one is 1/4" size aggregates and the finer one is the crushed fine. The crushed fine is a secondary crushing product from the 1/4" aggregates. Therefore, it can be well assumed that both rock materials have a same mineral and chemical properties. Original gradation and specific gravity were determined firstly. The values list in Table 1. The asphalt binder used is from San Joaquin refinery with regular PG grade 64–10.

Table 1. Gradation and specific gravity for 1/4" and crushed fine aggregates

	1/4" size aggregates	Crushed Fine
Apparent specific gravity	2.819	2.786
Bulk specific gravity	2.675	2.650
Gradation		
#4	61.5	98.4
#8	6.9	80.5
#16	4.2	58.3
#30	3.4	43.1
#50	2.8	31.7
#100	2.3	22.4
#200	1.6	15.4

During the preparation of lab mixes, all aggregates were combined and split evenly and then oven-dried at 110 °C after delivered from the quarry. Aggregates used for asphalt content correction factor testing were split directly to certain smaller representative portions. While in order to obtain exact same gradation, aggregates used for gradation correction factor test were first sieved into different size fractions and then combined in batches by proportion. Aggregates were mixed with asphalt binder at 140 °C to make the mixture. All tests were replicated and conducted at least twice under the same condition. The Thermolyne® NCAT ignition oven furnace was selected for use in this study, which can automatically weigh, record and report completed test results. Two specified oven setting temperature –482 and 538 °C are used in this study.

4   Asphalt Content Correction Factor

For this test, HMA samples were prepared with four asphalt contents by weight of aggregate. Since finer materials have higher surface area and need more asphalt to thoroughly coat them, the crushed fine and 1/4" size aggregate used different asphalt content. The asphalt contents are 2.5, 3.5, 4.5, 5.0% for 1/4" aggregates and 5.0, 5.5, 6.5 and 7.5% for crushed fine. Figures 1, 2, 3, and 4 show the appearance of mixed materials in varied asphalt contents. Two different ignition oven temperatures –482 and 538 °C were used in this study specified in the AASHTO T308 test method.

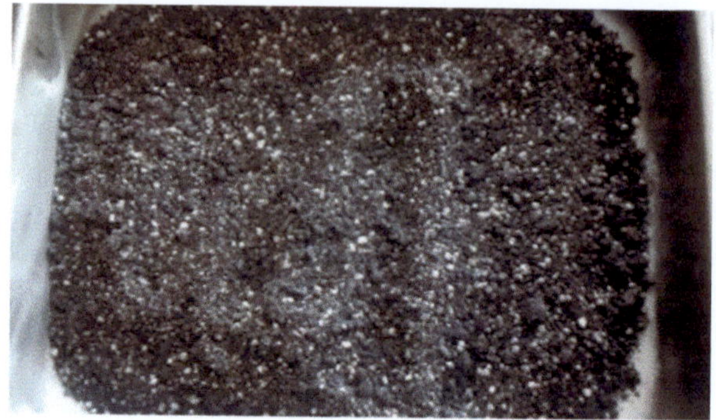

Fig. 1 5.5% asphalt content for crushed fine mixture

Fig. 2 7.5% asphalt content for crushed fine mixture

Fig. 3 2.5% asphalt content for 1/4" aggregate mixture

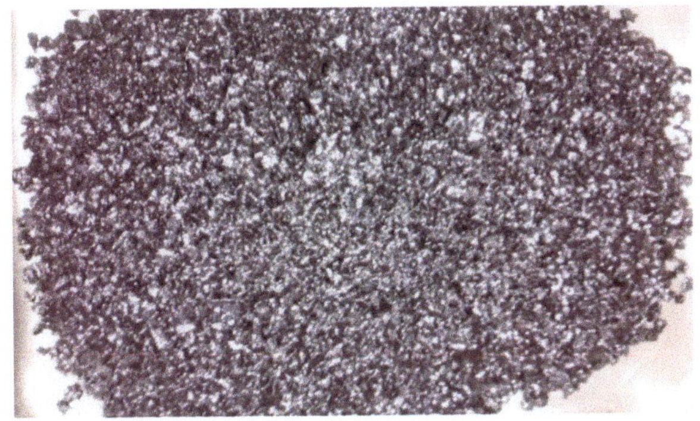

Fig. 4 5.0% asphalt content for 1/4" aggregate asphalt mixture

Correction factor was firstly investigated at 482 °C. Figure 5 shows the burning temperature results for 1/4" mix. It is observed that the peak temperature was reached about 10–11 minutes after placing the sample into the oven. The oven temperature then slowly decreased to the setting temperature from peak after about 30 min, by the time all of the binder has ignited. As shown in Fig. 5, temperature curves in each asphalt content are different. The mix with higher binder content remarkably shows higher peak temperature. The highest/peak temperatures range from 490 to 580 °C based on the lower to higher asphalt binder contents.

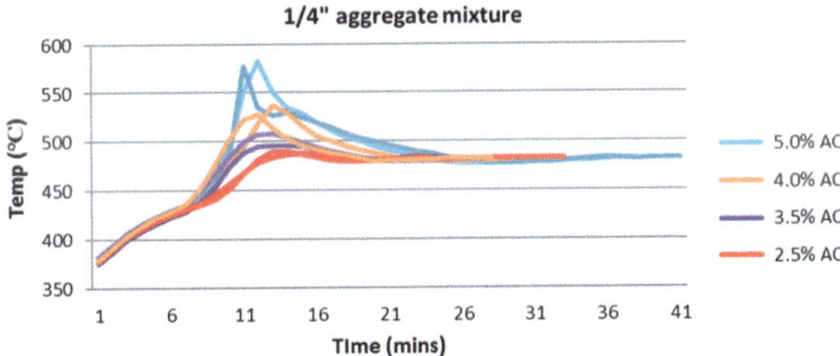

Fig. 5 Influence of asphalt content on 1/4" size mixture @ 482 °C

Figure 6 shows the burning temperature results for crushed fine materials. Temperature curves are similar to the 1/4" mix. Higher asphalt content leads to increased burning temperature. The highest/peak temperatures range from 520 to 680 °C for the four different asphalt contents. The burning peak temperatures are higher than the 1/4" size mix. This may indicate burning temperature is closely related to aggregate size and asphalt content. Figure 7 shows that the asphalt content correction factor for 1/4" and crushed fine mix. The results indicate that finer material (crushed fine) have higher correction factor than the larger size materials (1/4" size). A similar trend was noticed by Hall and Williams (1998) and Kowalski et al. (2010). The cause could be due to the higher peak temperature for smaller size aggregate. The peak burning temperature for crushed fine is about 50 °C higher than that of 1/4" size mix. Similarly, as for the same materials with different asphalt content, higher asphalt content leads to higher correction factor. The results suggest that the increased binder content may cause higher burning and peak temperature and somewhat more decomposition. It also indicates that the mass loss probably is due to decomposition of the aggregates not the removal of asphalt binder. Based on the findings, it could be addressed that both asphalt binder content and aggregate size have a compound effect on the ignition oven correction factor.

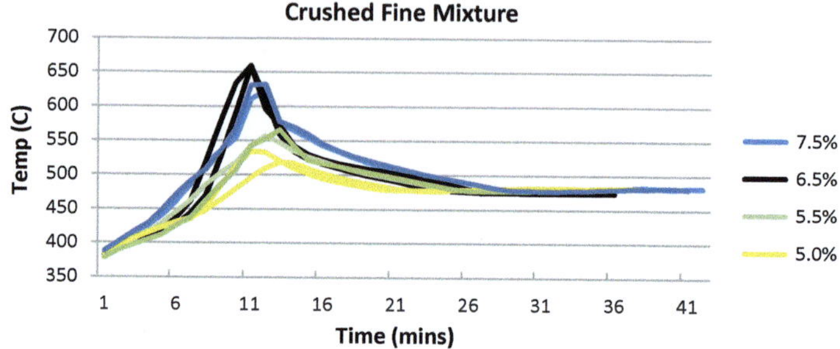

Fig. 6 Influence of asphalt content on crushed fine mixture @ 482 °C

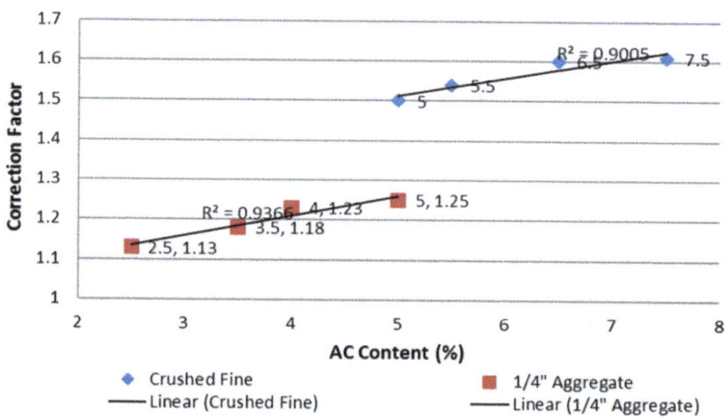

Fig. 7 Influence of asphalt content on correction factor @ 482 °C

The influence of ignition oven setting temperature was conducted only at 2.5 and 5.0% asphalt content for 1/4" aggregates and 5.0 and 7.5% for crushed fine. The temperature curves will be discussed later in the aggregate gradation correction factor section. The observed mass losses are clearly related to oven setting temperature, with higher temperature leading to higher loss. Specifically, the correction factor would reduce by about 0.1 from 538 to 482 °C as shown in Fig. 8.

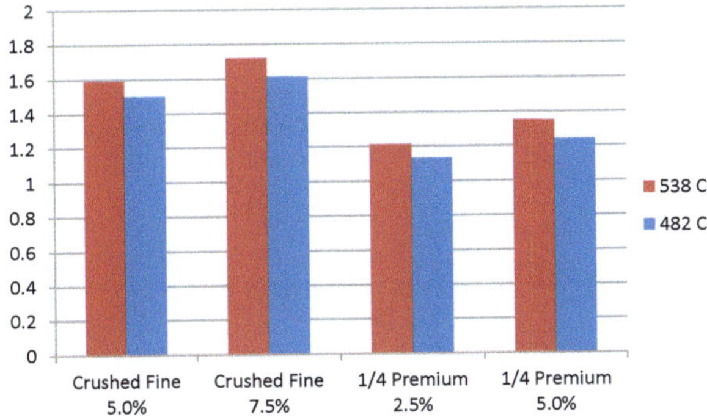

Fig. 8 Ignition oven correction factor at 482 and 538 °C

5   Aggregate Gradation Correction Factor

As a second objective of this study, the post-burn gradation has also been investigated. Figures 9 and 10 show the temperature curves for 1/4" size and crushed fine materials. The temperature curves look closer for each replication. This means batched materials would have a more uniform burning temperature environment than the split sample used in the previous asphalt content correction factor study. It also indicates that the ignition oven test could be sensitive to the gradation and mass variation. The figures also show that temperature curves look parallel with same asphalt content at 482 and 538 °C. This may suggest ignition oven setting temperature would increase the overall oven temperature but do not change the temperature trend during the test, which lead to a fixed increase rate of correction factor by only 0.1 for both crushed fine and 1/4" aggregates. Figures 9 and 10 also shows that the higher the test temperature, the sooner the oven temperature peaked. Specimens in the 538 °C oven all reached an earlier peak (2–3 mins earlier) temperature comparing to the test under 482 °C.

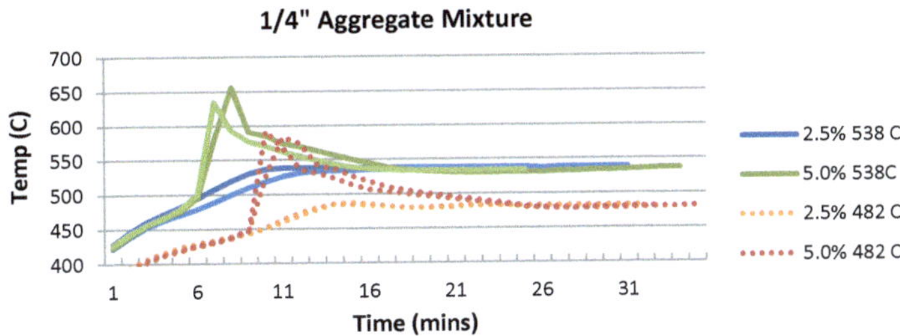

Fig. 9 Temperature plot for batched 1/4" aggregate mixture

Tables 2 and 3 list out the particle breakdown and gradation changes for the 1/4" size and crushed fine after the ignition oven tests, and the batched gradation before the test as listed in the "batched gradation" column. Only two different asphalt contents

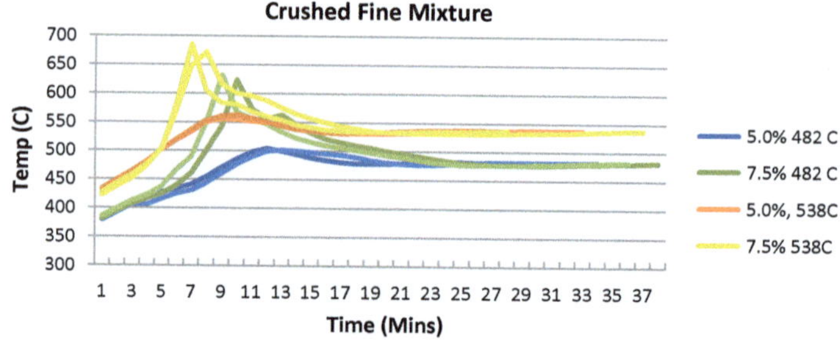

Fig. 10 Temperature plot for batched crushed fine mixture

were used for the evaluation of gradation correction factor. 2.5 and 5.0% asphalt content were used for 1/4" aggregates and 5.0 and 7.5% asphalt content were used for crushed fine materials. As can be seen, the post-burn gradation is skewed slightly to the finer side at a higher temperature and higher asphalt content for both materials. A larger gradation difference was observed for the 1/4" size at 538 °C, and obvious breakdown occurs on the No. 4 and No.8 sieves. Also, severe aggregate breakdown happens on the No. 4 and No. 8 sieves in the 482 °C oven setting temperature, however, the post-burn gradation difference between the two oven setting temperatures is not obvious.

Table 2. Summary for 1/4" aggregate mixture post-burn gradation

Sieve size	Batched gradation	5.0% AC @ 538 °C		2.5% AC @ 538 °C		5.0% AC @ 482 °C		2.5% AC @ 482 °C	
#4	67.5	73.7	73.9	72.7	72.7	72.6	73.3	72.9	73.0
#8	5.8	14.1	14.0	13.4	13.4	13.0	13.1	12.7	12.5
#16	4.5	10.4	10.4	9.9	9.9	9.8	9.8	9.6	9.4
#30	4.2	8.7	8.7	8.3	8.4	8.4	8.2	8.2	8.0
#50	3.9	7.4	7.4	7.1	7.1	7.2	7.0	7.1	7.0
#100	3.6	6.2	6.2	6.0	6.1	6.2	6.0	6.0	6.0
#200	3.1	4.9	4.9	4.8	4.9	5.0	5.0	4.9	4.9

For the crushed fine materials, no apparent gradation change or aggregate breakdown was observed at the two different asphalt contents and oven temperatures. The difference between pre-burn/ batched gradation and post-burn gradation is also very small. This is not expected, since crushed fine has higher asphalt content correction factor and was observed to have a higher burning temperature, which should lead to more breakdown. One of the reasons could be, the crushed fine, as the secondary product of 1/4" size aggregates with further times of crushing, may have higher resistance and less breakdown under high temperature or mechanical load. A similar phenomenon was also observed by Harrell et al. (2006) using dolomite. It was found that the degree of decrepitation increases with the size of the crystals within dolomite

Table 3. Summary for crushed fine mixture post-burn gradation

Sieve size	Batched gradation	7.5% AC @ 538 °C		5.0% AC @ 538 °C		7.5% AC @ 482 °C		5.0% AC @ 482 °C	
#4	98.3	98.5	98.6	98.4	98.7	99.0	98.8	98.3	98.6
#8	80.7	81.7	81.6	81.5	81.8	81.5	81.6	81.4	81.6
#16	57.9	61.4	60.2	60.4	60.7	60.1	60.2	57.9	58.8
#30	41.3	43.1	43.1	42.9	42.9	42.9	43.0	42.7	42.7
#50	28.7	31.0	31.2	30.5	30.7	30.8	30.4	30.2	30.1
#100	19.8	21.1	21.1	20.3	20.6	20.9	20.7	20.4	20.9
#200	12.8	12.6	12.1	11.7	11.9	12.6	12.6	11.9	12.1

particles. It is hypothesized that particles with larger crystals have longer pre-existing flaws and may be less retard to the propagation of fractures and higher breakdown.

6 Polarized Light Microscope Test

Mineral testing was run in this study. The test used thin-cut from the aggregate sample and analyzed under microscopy by polarized light. Three combined samples were prepared. The combined sample blends 50% crushed fines and 50% 1/4" aggregates.

- Combined Sample #1: drying in 110 °C ovens for one hour.
- Combined Sample #2: Combined aggregates were tested alone in 500 °C ignition oven for 50 mins (never exposed to the binder).
- Combined Sample #3: Combined aggregates were mixed with 5.0% of asphalt and burning in the ignition oven (temperature ranges from 482 to 600 °C) until all the asphalt binder was burnt out

High temperature would transform the minerals and bulk chemistry of granitic rock. Sample #1 was only dried at 110 °C oven. Image from the polarized light testing shows that larger grains are still intact. Aggregate mineral grains were well-formed. Clay/alteration seems to be present in the #1 sample as shown in Fig. 11. The portion of altered minerals by clay and chlorite is in violet and purple color. The alteration is likely caused by mild temperature drying at 110 °C. In Fig. 12, Hornblendes in shades of light green and grey were fractured and more breakdowns were noticed. Sections of altered aggregate were also noticed including brown fractured chlorite, and hornblende replaced by chlorite. For sample #3, minerals suffered further breakdown and more decomposition comparing to the sample #2. Ignition of the binder for this sample causes the oven temperature to increase to the peak temperature at around 600 °C. More plagioclase is replaced by feldspar in the sample in white color as can be seen in Fig. 13.

Fig. 11 Combined Sample #1

Fig. 12 Combined Sample #2

Fig. 13 Combined Sample #3

7 Conclusions

The ignition method is the most common way for estimating asphalt content of HMA while also obtaining a "clean" sample of aggregate for determining the aggregate blend gradation without requiring the use of environmentally hazardous chemicals or radioactive sources. Based on the work performed in this study using granite aggregates, some potentially conclusions can be offered

- Crushed fine material shows higher asphalt content correction factor than the 1/4" size aggregates. This may indicate us that smaller size aggregates have a higher asphalt content correction factor than the larger sized aggregates.
- Asphalt content correction factor changes with temperature. Higher temperature leads to a higher correction factor. For example, the correction factor at 538 °C is higher than that at 482 °C.
- Asphalt content is another factor on ignition oven burning temperature. The oil wrapped thickness/film thickness is dependent on asphalt contents. Thicker HMA film thickness would lead to higher burning peak temperature.
- Post-burn gradation shows that larger size aggregate (1/4" size) has a higher breakdown than the smaller size aggregate (Crushed fine).
- Higher asphalt content leads to higher burning peak temperature and more aggregate breakdown. However, the gradation change is not obvious for both larger and smaller aggregates.

- Burn gradation is not closely related to the burning temperature either. No obvious difference was seen for the post-burned gradation at 538 °C and 482 °C oven temperatures.
- The polarized light microscope analysis clearly demonstrates that granite aggregates that have been subjected to the ignition oven have partially decomposed, producing chlorite and hornblende. It also shows that the presence of the binder does increase the extent of dolomite decomposition in the ignition oven.

References

AASHTO: Determining the asphalt binder content of hot mix asphalt by the ignition method – AASHTO Designation T308-10 (2014)

Dollimore, D., Dunn, J.G., Lee, Y.F., Penrod, B.M.: The decrepitation of dolomite and limestone. Thermochim. Acta **237**(1), 125–131 (1994)

Hall, K.D., Williams, S.G.: Effects of the ignition method on aggregate properties. Paper prepared for publication and presentation at the 1999 annual meeting of the association of asphalt paving technologists (1998)

Harrell, J.A., Dunn, J.G., Welshimer, J.W.: Effect of crystal size on dolomite decrepitation in glass furnaces. Glass Technol. Eur. J. Glass Sci. Technol. **47**(6), 188–192 (2006)

Kowalski, K.J., McDaniel, R.S. Olek, J., Shah, A. (2010). Determining of the binder content of hot mix asphalt containing dolomitic aggregates using the ignition oven. Joint Transportation Research Program, FHWA/IN/JTRP-2010/13

McCauley, R.A., Johnson, L.A.: Decrepitation and thermal decomposition of dolomite. Thermochim. Acta **185**(2), 271–282 (1991). https://doi.org/10.1016/0040-6031(91)80049-O

Smith, F.G.: Decrepitation characteristics of igneous rocks. The Canadian Mineralogist (1957)

Author Index

© Springer International Publishing AG, part of Springer Nature 2019
W. Frikha et al. (eds.), *New Developments in Soil Characterization and Soil Stability*,
Sustainable Civil Infrastructures, https://doi.org/10.1007/978-3-319-95756-2